Medical and Biologic Effects of Environmental Pollutants

PLATINUM-GROUP METALS

Subcommittee on Platinum-Group Metals

Committee on Medical and Biologic Effects of Environmental Pollutants

DIVISION OF MEDICAL SCIENCES
ASSEMBLY OF LIFE SCIENCES
NATIONAL RESEARCH COUNCIL

NATIONAL ACADEMY OF SCIENCES
WASHINGTON, D.C.   1977

Other volumes in the Medical and Biologic Effects of Environmental Pollutants series (formerly named Biologic Effects of Atmospheric Pollutants):

ARSENIC (ISBN 0-309-02604-0)
ASBESTOS (ISBN 0-309-01927-3)
CARBON MONOXIDE (ISBN 0-309-02631-8)
CHLORINE AND HYDROGEN CHLORIDE (ISBN 0-309-02519-2)
CHROMIUM (ISBN 0-309-02217-7)
COPPER (ISBN 0-309-02536-2)
FLUORIDES (ISBN 0-309-01922-2)
LEAD (ISBN 0-309-01941-9)
MANGANESE (ISBN 0-309-02143-X)
NICKEL (ISBN 0-309-02314-9)
NITROGEN OXIDES (ISBN 0-309-02615-6)
OZONE AND OTHER PHOTOCHEMICAL OXIDANTS (ISBN 0-309-02531-1)
PARTICULATE POLYCYCLIC ORGANIC MATTER (ISBN 0-309-02027-1)
SELENIUM (ISBN 0-309-02503-6)
VANADIUM (ISBN 0-309-02218-5)
VAPOR-PHASE ORGANIC POLLUTANTS (ISBN 0-309-02441-2)

NOTICE: The project that is the subject of this report was approved by the Governing Board of the National Research Council, whose members are drawn from the Councils of the National Academy of Sciences, the National Academy of Engineering, and the Institute of Medicine. The members of the Committee responsible for the report were chosen for their special competences and with regard for appropriate balance.

This report has been reviewed by a group other than the authors according to procedures approved by a Report Review Committee consisting of members of the National Academy of Sciences, the National Academy of Engineering, and the Institute of Medicine.

The work on which this publication is based was performed pursuant to Contract No. 68-02-1226 with the Environmental Protection Agency.

Library of Congress Catalog Card Number 77-90130

International Standard Book Number 0-309-02640-7

Available from
Printing and Publishing Office, National Academy of Sciences,
2101 Constitution Avenue, N.W., Washington, D.C. 20418

Printed in the United States of America

Preface

Since the spring of 1970, the Division of Medical Sciences, National Research Council, has produced several documents on the medical, biologic, and environmental effects of selected pollutants. These documents have been prepared for the Environmental Protection Agency to establish a broad background of information, to evaluate that information, and to recommend studies aimed at remedying information inadequacies or gaps. The documents are prepared by subcommittees of the Committee on Medical and Biologic Effects of Environmental Pollutants. This report is the result of the work of the Subcommittee on Platinum-Group Metals.

The purpose of this document is to assemble, organize, and evaluate all pertinent information (up to April 1976) about the effects on man and his environment that result either directly or indirectly from pollution by platinum-group metals: iridium, osmium, palladium, platinum, rhodium, and ruthenium. The document describes physical and chemical properties, sources, measurement, and effects of plants, animals, and humans. The information presented is supported by references to the scientific literature, and the summary, conclusions, and recommendations represent a consensus of the members of the Subcommittee.

SUBCOMMITTEE ON PLATINUM-GROUP METALS

JOE W. HIGHTOWER, Rice University, Houston, Texas, *Chairman*
RICHARD H. ADAMSON, National Institutes of Health, Bethesda, Maryland
JOHN L. BEAR, University of Houston, Houston, Texas
HENRY F. FREISER, University of Arizona, Tucson, Arizona
VLADIMIR HAENSEL, Universal Oil Products Company, Des Plaines, Illinois
WILLIAM A. E. MCBRYDE, University of Waterloo, Waterloo, Ontario, Canada
BARNETT ROSENBERG, Michigan State University, East Lansing, Michigan

J. PEPYS, Cardiothoracic Institute, London, England, *Consultant*

JAMES A. FRAZIER, National Research Council, Washington, D.C., *Staff Officer*

COMMITTEE ON MEDICAL AND BIOLOGIC EFFECTS OF ENVIRONMENTAL POLLUTANTS

HERSCHEL E. GRIFFIN, University of Pittsburgh, Pittsburgh, Pennsylvania, *Chairman*
MARTIN ALEXANDER, Cornell University, Ithaca, New York
ANDREW A. BENSON, University of California, La Jolla, California
RONALD F. COBURN, University of Pennsylvania School of Medicine, Philadelphia, Pennsylvania
T. TIMOTHY CROCKER, University of California College of Medicine, Irvine, California
CLEMENT A. FINCH, University of Washington School of Medicine, Seattle, Washington
EVILLE GORHAM, University of Minnesota, Minneapolis, Minnesota
ROBERT I. HENKIN, Georgetown University Medical Center, Washington, D.C.
IAN T. T. HIGGINS, University of Michigan, Ann Arbor, Michigan
JOE W. HIGHTOWER, Rice University, Houston, Texas
HENRY KAMIN, Duke University Medical Center, Durham, North Carolina
ORVILLE A. LEVANDER, Agricultural Research Center, Beltsville, Maryland
I. HERBERT SCHEINBERG, Albert Einstein College of Medicine, Bronx, New York
ROGER P. SMITH, Dartmouth Medical School, Hanover, New Hampshire

T. D. BOAZ, JR., National Research Council, Washington, D.C., *Executive Director*

Acknowledgments

This document was prepared by the members of the Subcommittee on Platinum-Group Metals. Each chapter was written by one or more members expert in the subject matter. The summaries, conclusions, and recommendations are the consensus of the Subcommittee members.

The Chairman of the Subcommittee was Dr. Joe W. Hightower, who wrote the Introduction and Chapter 8 with Dr. Vladimir Haensel. Dr. William A. E. McBryde wrote Chapter 2 and was jointly responsible for Chapter 3 with Drs. Richard H. Adamson and Barnett Rosenberg, for Chapter 4 with Dr. John L. Bear, and for Chapter 5 with Dr. Henry Freiser. Chapter 6 was prepared jointly by Drs. Adamson and Rosenberg. Dr. Jack Pepys wrote Chapter 7, and Dr. Haensel prepared the information that appears in the Appendix.

The preparation of the report was assisted by the comments of anonymous reviewers chosen by Dr. Ralph G. Smith, who served as Associate Editor. The members of the Committee on Medical and Biologic Effects of Environmental Pollutants (MBEEP) were very helpful in reviewing and commenting on the report. In addition, several liaison representatives to the MBEEP Committee, both inside and outside the National Academy of Sciences–National Research Council, provided helpful comments.

Information resources and various documents were obtained from Dr. R. J. M. Horton, Mr. M. Malanchuk, Dr. Wellington Moore, and Mr. John B. Moran of the Environmental Protection Agency. Informational assistance was given by representatives of American Cyanamide Company; American Oil Company; Engelhard Minerals and Chemicals Corporation; Exxon Research and Engineering Company; Ford Motor Company; General Motors Technical Center; Matthey Bishop, Inc.; and Union Oil of California. Additional assistance was obtained from the National Research Council's Advisory Center on Toxicology, the National Academy of Sciences Library, the Library of Congress, the National Library of Medicine, and the Air Pollution Technical Information Center.

The staff officer for the Subcommittee was Mr. James A. Frazier. The editor was Mr. Norman Grossblatt, and the reference assistant was Ms. Joan Stokes. The report was typed by Mrs. Eileen Brown.

Contents

1	Introduction	1
2	Sources and Production	3
3	Uses	14
4	Physical and Chemical Properties	39
5	Analysis and Determination	66
6	Toxicology and Pharmacology	79
7	Allergy to Platinum Compounds	105
8	Environmental Considerations	125
9	Summary	165
10	Conclusions	173
11	Recommendations	177
	Appendix: Catalytic Reforming in Refining of Petroleum	181
	References	199
	Index	225

1

Introduction

Six elements of group VIII (in the periodic table) have been collectively designated the "platinum-group metals." Included in this group are platinum (Pt), palladium (Pd), rhodium (Rh), ruthenium (Ru), iridium (Ir), and osmium (Os). Sometimes called "noble metals" (with gold, silver, etc.), because of their resistance to oxidation, these "precious metals" are present in the earth's crust in very low concentrations. In spite of their limited availability, these metals (and chemical compounds containing them) are extremely useful as catalysts in the chemical and petroleum industries, as conductors in the electric industry, in extrusion devices, in dental and medical prostheses, and in jewelry.

In their traditional applications, the platinum-group metals have been considered relatively innocuous, with respect to direct environmental impact. However, some new and more extensive uses of these materials may have both direct and indirect impact on human health. It has been known for years that physiologic activity is associated with some platinum-group metal compounds. Some of the complex salts produce allergic reactions (e.g., platinosis) in humans, and a few of the volatile oxides are very toxic. This report presents information about the health effects of the platinum-group metals and their compounds and identifies the hazards associated with them.

Since the beginning of the 1975 model year, most new automobiles sold in the United States have been equipped with catalytic converters whose purpose is the chemical removal of some polluting substances from the exhaust. The amount of platinum-group metals required in these devices is about equal to the total amount in all other U.S. uses combined. Concern has been expressed about the possibility that harmful compounds of these metals will be directly emitted from the exhaust systems. In addition, the catalytic converters are responsible for emission of sulfuric acid, and it is possible that the acid will reach harmful concentrations in the vicinity of heavily traveled highways. Both direct metal emission and indirect sulfuric acid emission are considered in this report.

Another possible application of platinum-group metals is the use of some complexes as antitumor agents in cancer chemotherapy. Recent successes in this field have stimulated intensive research to find compounds that are highly effective either by themselves or in combination with other drugs in retarding cancer, but that have minimal harmful side effects on such organs as the kidneys.

An unusual mode of entry of some of the platinum-group metals into the environment is through waste effluent from nuclear-fuel reprocessing plants.

In assessing the environmental impact of the platinum-group metals and their compounds, this report considers the sources and uses of these materials, their physicochemical properties and associated analytic methods, their toxicology and pharmacology, and problems involved in the refinement and preparation of end products containing them. For further information on the chemistry of these metals, the reader is referred to a number of important books.[78,109,170,186,235,258,263,264,485]

2

Sources and Production

GEOLOGIC OCCURRENCE[113,490]

In most of their natural occurrences, the platinum-group metals are either uncombined or "native"; only a few well-characterized mineral compounds are known. For instance, native platinum generally contains small concentrations of the other platinum metals (except osmium), base metals (such as iron, copper, and nickel), or silver. Platinum is also a major or minor component of the other native alloys of the platinum metals listed in Table 2–1.

The most familiar mineral species of these metals are sperrylite, $PtAs_2$; cooperite, $(Pt,Pd)S$; braggite, $(Pt,Pd,Ni)S$; potarite, $PdHg$ or Pd_3Hg_2; stibiopalladinite, Pd_3Sb; and laurite, $(Ru,Os)S_2$. In recent years, a number of other mineral species have been identified, mainly as a result of the application of new physical techniques, such as x-ray spectroscopy and electron-probe analysis.

The abundance of the six platinum metals in the earth's crust appears to be very low (see Table 2–2); in fact, they are between seventy-first and eightieth among all the elements. However, they are much more abundant in meteorites, particularly meteorites or phases of meteorites that are rich in metals, notably iron and nickel, as opposed to the "stony" meteorites. In geochemical terminology, the platinum metals are said to be strongly siderophile (in contrast with chalcophile). This has led geochemists to the supposition that the platinum metals have

TABLE 2-1 Native Platinum Metals and Alloys[a]

Name	Crystal Form	Principal Constituents
Platinum	Cubic	Pt
Platiniridium	Cubic	Pt, Ir
Palladiplatinum	Cubic	Pd, Pt
Palladium	Cubic	Pd
Allopalladium	Hexagonal	Pd
Iridosmine	Hexagonal	Ir, Os
Osmiridium	Cubic	Os, Ir

[a] Data from Wright and Fleischer[490] and Crockett.[113]

TABLE 2-2 Abundance of Platinum Metals in Earth's Crust[a]

Metal	Concentration in Crust, ppm
Palladium	0.01
Platinum	0.005
Rhodium	0.001
Iridium	0.001
Ruthenium	0.001
Osmium	0.001

[a] Data from Mason.[269]

concentrated during the earth's formation mainly in the iron-nickel core and that this accounts for their relatively low abundance in the lithosphere, or rocky crust of the earth.[163] A similar preferential distribution occurs during the smelting of ores and results in a concentration of platinum metals when conditions are chosen so as to produce fused base metals (e.g., iron, nickel, and copper) and liquid slag; the precious metals are found almost entirely in the metallic phase.

The platinum-group metals have been found in minerals in the earth's crust at concentrations up to 20 ppm (1 ppm = 10^{-4} wt%)—but usually much lower—and most commonly in sulfides, selenides, tellurides, and arsenides. Similar abundances have been noted in some oxide minerals, especially chromite. Unfortunately, at such low concentrations, it has not yet proved possible to state in what chemical forms the platinum metals are present (e.g., as distinct mineral species or in solid solutions). Current data reveal that, when these metals are

Sources and Production

present in rocks, there is a marked preferential association with ultrabasic, rather than silicic, species. The former are believed to have crystallized early in the sequence of magmatic separation, and it has been proposed[352] that the comparatively unreactive and high-melting-point platinum metals would have been deposited early and become associated with the early fractions crystallizing from the magma. Accordingly, these metals appear in concentrations up to 1 ppm in dunites, pyroxinites, and serpentines, but in very low concentrations in rocks that contain more silica and that are regarded as having crystallized later.

SOURCES FOR COMMERCIAL PRODUCTION

Platinum metals available for industry and commerce are either extracted from newly mined mineral (as described in the next section) or obtained by refining used scrap metal. Because of the high intrinsic value of the metals, coupled with their chemical inertness and physical durability, used metal can profitably be recycled; a significant fraction of annual production includes such metal. Reclaimed metals are often referred to as having been "toll-refined."*

The economically significant sources of platinum metals are in the Republic of South Africa, Canada, and the U.S.S.R. They are all primary deposits usually associated with ultrabasic rock formations and often with copper and nickel sulfide deposits. The South African sources are the most concentrated, but even they contain the platinum metals at only 4–10 ppm. These sources are in the Bushveld Igneous Complex in the Transvaal district just north of Pretoria.[112] In Canada, the platinum metals occur mainly in copper–nickel sulfide ores in the Sudbury area of Ontario and in the Thompson–Wabowden area of Manitoba.[75,190] In the Sudbury deposits, the platinum-metal content has been estimated at less than 1 ppm, but the metals are concentrated to profitable values during the refining of copper and nickel. In the U.S.S.R., the richest sources are in the Noril'sk region of Siberia and in the Kola Peninsula near Petsamo. The abundance (concentration) of platinum metals in the Russian deposits has not been disclosed, but is believed to be between those found in Canada and South Africa.

Almost 90% of the platinum metals in these sources consists of platinum and palladium. Table 2–3 shows the relative abundance by weight of the various metals in the three principal sources.

---

*Because of the specialized technology involved, users and some producers of the platinum metals send their material to refiners for separation, purification, etc., for which service a fee or toll is charged; hence, "toll-refining."

TABLE 2-3 Estimated Composition of Platinum-Group Metals from Different Sources[a]

Metal	Percentage by Weight		
	Canada	U.S.S.R.	South Africa
Platinum	43.4	30	64.02
Palladium	42.9	60	25.61
Iridium	2.2	2	0.64
Rhodium	3.0	2	3.20
Ruthenium	8.5	6	6.40
Osmium	0	0	0.13
Approximate ratio Pt:Pd	1:1	1:2	2.5:1

[a]Data from Mitko[280] and *Platinum Metals Review*.[22]

The first platinum to be identified came from South America and was brought to Europe in the seventeenth and eighteenth centuries as an almost undesirable adjunct of gold from the New World. In fact, the name "platina," first applied to this material by the Spaniards, signifies "little silver"—a debased form, perhaps, of silver. It was from placer deposits,* some of which in Colombia are still being mined. Smaller quantities of placer platinum are also being obtained from western Alaska, Ethiopia, and the Philippines. The total amount of new metal from all these sources is relatively small. In the nineteenth century, extensive placer deposits in the Ural Mountains region in Russia constituted the chief world source of platinum metals, and a small amount is still obtained from that location. The metal in these placer deposits is present as native alloys of varied composition.

The most recent available information on worldwide production of new platinum metals is incorporated in Table 2-4.

Of platinum metals refined in the United States in 1973 (the latest available data),[14] new metal, either placer or byproduct from gold- and copper-refining, accounted for 8,218 troy oz (platinum, 43%; palladium, 42%), whereas secondary metal recovered by recycling amounted to 232,276 troy oz (platinum, 36.8%; palladium, 55.4%). These numbers, especially those for new metal, tend to fluctuate somewhat from year to year. Another statistic pertaining to refining of these metals in the United States is the amount handled "on toll," a

---

*The term "placer" signifies a deposit of sand or gravel of alluvial or glacial origin and containing particles of gold or other precious minerals. Normally, the metal, being heavier, is separated by washing away the sand or gravel ("panning").

total of 1,361,723 troy oz in 1972, of which 84,219 oz were of crude or matte from Colombia, Canada, and South Africa.

It is impossible to determine the exact amounts of the platinum metals that can be recovered by mining. Estimates of the known world reserves of platinum and palladium are shown in Table 2–5.

The Johns–Manville Corporation[231,232] recently reported finding a large deposit of ore rich in platinum and palladium in the Stillwater Complex area, Sweetgrass County, Montana. Although complete information is not available, estimates of the recoverable metals run as high as 500 million troy ounces—comparable with the total world reserves previously thought to exist. The ore is highly concentrated, in the range of 10–12 pennyweight/ton,* or about 10 ppm. The ore is rich in palladium, the platinum:palladium ratio being about 1:3.5. This is the most extensive deposit and most concentrated ore yet reported within the United States, and studies are under way to determine the economic feasibility of mining it commercially.

PRODUCTION AND REFINEMENT

There are two principal stages in the isolation of reasonably pure platinum metals from raw materials. One is extraction of a concentrate of precious metals from a large body of ore. The other is the refining of the precious metals, which involves their separation from the concentrate and from each other and ultimately their purification. As indicated earlier, refining applies not only to native sources of new metal, but also to a large quantity of scrap and other used metal that is recycled. The details of the first of these operations depend on the source of the raw material and its composition.

Rather complete descriptions of the metallurgic operations of the International Nickel Company in Sudbury, Ontario,[121] and of the Rustenberg Platinum Mines Limited in South Africa[38] are available, and what follows is only a brief synopsis.

In the International Nickel process, most of the platinum-group metal is separated from the bulk of the copper and nickel during slow cooling of a Bessemer matte. During this cooling, the oxidation of sulfur is regulated so as to produce small amounts of metallic nickel and copper. The latter serve as collectors of the precious metals from the original ore, and separation of the metallic phase is facilitated, because this phase is magnetic. The separated material can be concentrated to an even richer alloy, the electrolytic refining of which yields a

---

*One pennyweight = 0.05 troy oz or 1.555 g.

TABLE 2–4 Approximate World Production of Platinum-Group Metals, 1971–1975[a]

Country and Metals	Quantity, troy oz[b]				
	1971	1972	1973	1974	1975
Australia:					
Palladium metal content (nickel ore)	—	—	750	860	1,400
Platinum metal content (nickel ore)	—	—	225	260	420
Canada:					
Platinum and platinum-group metals (nickel ore)	475,169	406,048	354,223	384,618	430,000
Colombia:					
Placer platinum	25,610	24,111	26,358	21,094	22,114
Ethiopia:					
Placer platinum	217	248	235	230	162
Finland:					
Platinum-group metals recovered (copper ore)	600	650	725	650	600

Japan:					
Palladium from nickel and copper refineries	5,375	5,659	5,834	11,104	13,981
Platinum from nickel and copper refineries	3,451	4,240	4,363	4,101	5,482
Philippines:					
Palladium metal	1,756	4,810	4,180	2,315	836
Platinum metal	703	2,712	2,476	1,350	579
Republic of South Africa:					
Platinum-group metals from platinum ores	1,250,000	1,450,000	2,360,000	2,832,000	2,620,000
Osmium–indium from gold ores	3,200	3,000	2,800	2,500	2,400
U.S.S.R.:					
Placer platinum and platinum-group metals from platinum–nickel–copper ores	2,300,000	2,350,000	2,450,000	2,500,000	2,650,000
United States:					
Crude placer platinum and byproduct from gold- or copper-refining	18,029	17,112	19,980	12,657	18,920
TOTAL	4,084,110	4,268,590	5,232,149	5,773,739	5,766,894

[a]Data from Butterman.[72,73]
[b]Precious metals in commerce and production are for traditional reasons measured in troy ounces; 1 troy ounce = 1.097 avoirdupois ounce = 0.0311 kg.

TABLE 2-5 Estimated World Reserves of Platinum and Palladium[a]

Country	Reserves, 1,000 troy oz		
	Platinum	Palladium	Total
South Africa	142,400	50,200	192,600
U.S.S.R.	60,000	120,000	180,000
Canada	6,940	6,860	13,800
Colombia	5,000	0	5,000
United States	950	1,960	2,910
TOTAL	215,290	179,020	394,310

[a]Excluding recent find in Montana. Data from U.S. Bureau of Mines.[450]

rich concentrate in the anodic slimes. Smaller amounts of the precious metals are also recovered during refining of nickel either electrolytically or by the Mond carbonyl process. The electrolytic refining operations of International Nickel are carried out at Port Colborne in the Niagara Peninsula in Ontario. Most of the osmium in the anodic sludges (or slime) is recovered and refined in the course of acid treatment of the roasted sludges. The remainder is sent to the refineries of the Mond Nickel Company, Acton (London), England.

Another company, Falconbridge Nickel Mines, Limited, operating in the Sudbury, Canada, region, follows a somewhat different treatment of the ore that results in a high-grade sulfide matte, rich in nickel and copper, which is shipped to Kristiansand, Norway, for further treatment. Electrolytic refining of the nickel in Norway also produces anodic sludges that can be worked up to a rich concentrate of precious metals, which is partially returned to North America for refining by Engelhard Minerals and Chemicals Corporation in New Jersey.

The South African ore is so rich in platinum metals that as much as two-thirds can be recovered by gravity concentration alone. The remainder can be obtained after smelting of the tailings, which contain appreciable amounts of copper and nickel. The process results in recovery of these metals; in the course of nickel-refining, platinum metals again accumulate in the electrolytic process. About three-fourths of the various South African concentrates of precious metals are shipped to England and refined at Brimsdown or at Royston, near London, by Johnson-Matthey Limited. The remaining one-fourth is processed by the Impala Company in South Africa.

The electrolytic anodic slimes and other rich concentrates are subjected to roasting and then acid-leaching to extract copper, nickel, or other base metals. Under such conditions, the osmium may form a volatile tetroxide (see Chapter 4) and be all or partially lost. Recently, the International Nickel Company gave details of a process for the easy recovery of most of the available osmium at this stage.[225] In this process, treatment of the dried anodic sludges with sulfuric acid is limited to temperatures not in excess of 200° C. This renders copper and nickel soluble, but leaves the platinum metals insoluble, except for a small amount (less than 5%) of osmium that is volatilized. This insoluble residue is dried and then brought to ignition at 800–900° C; this volatilizes about 85% of the osmium as the tetroxide, OsO_4. The tetroxide is absorbed in alkaline scrubbing solution and retained for further purification.

The ignited acid-soluble residue is freed of sulfur, selenium, and arsenic by this treatment.[36] It is then treated with aqua regia, which dissolves most of the platinum, palladium, and gold. From the resulting solution, gold is precipitated by addition of a ferrous salt. After separation of the gold, ammonium chloride is added to precipitate the yellow salt ammonium hexachloroplatinate, $(NH_4)_2PtCl_6$; this is separated and ignited to yield the metal. For further purification of the platinum, it is redissolved in aqua regia, and the solution is evaporated gently with sodium chloride and hydrochloric acid to destroy oxides of nitrogen and nitrosyl compounds. This solution of sodium hexachloroplatinate(IV), Na_2PtCl_6, is then treated with sodium bromate and its pH carefully raised to cause precipitation of the hydrous oxides of any rhodium, iridium, or palladium that was carried down in the initial precipitation of platinum. After removal of solids by filtration, the solution is boiled with hydrochloric acid to destroy bromate, and then treated with ammonium chloride for a second precipitation of platinum as ammonium hexachloroplatinate(IV). This is filtered off, dried, and brought slowly to ignition at 1000° C. The resulting product is a gray sponge of metal in a pure (over 99.9%) form.

The filtrate after removal of gold and platinum contains palladium as chloropalladous acid, H_2PdCl_4. Addition of aqueous ammonia to this solution causes precipitation of the yellow complex diammine dichloride, $Pd(NH_3)_2Cl_2$, which redissolves in excess ammonia through formation of the complex ion $Pd(NH_3)_4^{2+}$. From the ammoniacal solution, the insoluble diammine dichloride is reprecipitated by the addition of hydrochloric acid. Palladium is isolated and further purified by successive precipitation and redissolution of this compound in this way. Ignition of the complex salt yields palladium metal in spongy form; the ignited sponge is usually cooled in an oxygen-free atmosphere to avoid superficial oxidation.

The residue not already dissolved must then be treated to recover silver, rhodium, iridium, ruthenium, and osmium. The precise details of the practices followed by different refiners in handling this material are not always disclosed and are certainly not uniform. In one account,[100] the insoluble residue is heated with fluxes, much as in the classic fire assay, to produce a lead alloy containing the precious metals. This is treated with nitric acid to dissolve the lead and silver. The insoluble residue is then fused with sodium bisulfate; this treatment selectively converts rhodium to a water-soluble sulfate. The solution produced is made alkaline to precipitate rhodium hydroxide, $Rh(OH)_3$, which is separated and dissolved in hydrochloric acid. Impurities are carried through this step with the rhodium, but may be separated later by hydrolytic precipitation in the presence of nitrite. The nitrite complex of rhodium, $Rh(NO_2)_6^{3-}$, is very stable over a wide range of pH and remains in solution under conditions suitable for precipitation of many metal hydroxides. The metal is later precipitated as ammonium hexanitrorhodate(III), $(NH_4)_3Rh(NO_2)_6$. Still further purification is achieved by converting the latter compound to the chlororhodite, $RhCl_6^{3-}$, through digestion with hydrochloric acid; the solution containing chlororhodite is passed through a cation-exchange column to remove traces of base metals. Rhodium is finally precipitated from this purified solution with formic acid, dried, and ignited under hydrogen to a residue of very pure metallic sponge.

The remaining metallic material is heated with sodium peroxide at 500° C to convert ruthenium and osmium (if this has not previously been recovered) to the water-soluble salts sodium ruthenate, Na_2RuO_4, and sodium osmate, Na_2OsO_4. Solutions of these compounds are then treated in such a way as to distill the volatile tetroxides out of these two metals. The treatment varies somewhat in accordance with the proportions of the two elements, but they are separated at this stage by taking advantage of the fact that osmium tetroxide, OsO_4, can be distilled from nitric acid solutions while ruthenium is retained in the pot as nitrosyl complexes. Distilled osmium tetroxide is collected in a solution of sodium hydroxide, usually containing alcohol. The absorbate is then digested with ammonium chloride, during which process osmyltetrammine chloride, $OsO_2(NH_3)_4Cl_2$, precipitates. This compound is dried and ignited in hydrogen to form a sponge of the pure metal. Other absorbing solutions and precipitation forms for osmium are also used. If osmium is absent or present at a low concentration, the solution extracted from the alkaline fusion is treated with chlorine gas and heated. Ruthenium distills as tetroxide under these conditions and is then absorbed in hydrochloric acid. Any osmium present at this stage may be distilled out by boiling with nitric acid. The remaining solution is boiled with hydrochloric acid to destroy nitrosyl complexes, and then the ruthenium is precipitated by the addition of ammonium chloride as ammonium hexachlororuthenate(III), $(NH_4)_3RuCl_6$. This may be heated in an inert atmosphere to yield the metal. If larger amounts of osmium are present, the amount of nitric acid required for its removal as tetroxide may result in an alternative precipitation form for ruthenium, namely, pentachloronitrosylruthenic acid(III), H_2RuCl_5NO.

The previously mentioned fusion with sodium peroxide converts iridium to its dioxide, which is insoluble in water. This may be brought into solution by treatment with aqua regia; from this solution, ammonium hexachloroiridate, $(NH_4)_2IrCl_6$, is precipitated by the addition of ammonium chloride and nitric acid. This salt may be further purified by dissolving it in ammonium sulfide solution, in which iridium remains soluble while impurities are precipitated. The latter are separated by filtration; ammonium chloroiridate is reprecipitated as before, then ignited in an atmosphere of hydrogen to yield a pure sponge of the metal.

A summary of the extraction and refining of the platinum metals from the principal South African source has been published elsewhere.[166] More recently, solvent extraction and ion-exchange techniques[130] for the recovery of the platinum-group metals have been tested to some extent, although their development has apparently reached only the pilot-plant stage.

3

Uses

In general, the uses of the platinum-group metals and their compounds derive from the special properties of these substances. The metals, alone or in alloys, have long been known to have extensive catalytic properties, the first such observations having been communicated by Sir Humphry Davy to the Royal Society in 1817. In recent years, a number of compounds of the platinum-group elements have also been introduced as catalysts in synthetic organic chemistry. Other properties that lead to the usefulness of these metals include their resistance to oxidation, even at high temperatures; resistance to corrosion; high melting point; high mechanical strength; and, at least for platinum and palladium, good ductility.

It may be broadly indicative of the nature and scale of the uses of the platinum-group metals to record their sales to consuming industries. Table 3–1 gives data for 1973 according to the American Bureau of Metal Statistics.[14] Tables 3–2 and 3–3 give data for platinum and palladium, according to the Bureau of Mines,[213,279,280,472] spanning a period of several years, from which any fluctuations and trends may be discerned. All data refer to the United States only.

CATALYTIC USES

The number of catalytic applications, their variety, and their selectivity for specific reactions make it impossible to do justice to the full range

Uses

TABLE 3-1 Sales of Platinum-Group Metals to Consuming Industries, United States, 1973[a]

Industrial Category	Quantity Sold, troy oz			
	Platinum	Palladium	Others	Total
Chemical	238,809	239,394	72,708	550,911
Petroleum	119,875	3,803	15,284	138,962
Glass	72,533	1,439	16,822	90,794
Electric	100,607	458,652	15,230	574,489
Dental and medical	18,103	62,776	1,153	82,032
Jewelry and decorative	20,366	21,658	13,698	55,722
Miscellaneous	53,479	61,254	11,095	125,828
TOTAL	623,772	848,976	145,990	1,618,738

[a] Data from American Bureau of Metal Statistics.[14]

of these applications in such a publication as this. In the survey that follows, a number of key references are cited, and these should be consulted for more details.

Naphtha-Reforming

The use of platinum in naphtha-reforming, introduced in 1949, has grown rapidly and today is one of its major industrial uses. The metal is dispersed on small pellets of alumina or silica–alumina, which, being porous themselves, expose an enormous specific surface area of the platinum. The petroleum fraction fed to the reactors comprises hydrocarbons, which boil at roughly 100–200° C, and hydrogen generated in the process. The reactions increase the octane rating of the gasoline fraction and produce large amounts of aromatic hydrocarbons that may be separated from the product and used for purposes other than fuel. Briefly, the reactions that take place during petroleum reformation include:

1. Conversion of naphthenes to aromatic hydrocarbons.
2. Isomerization or cracking of paraffin hydrocarbons and conversion of some of them to aromatics.
3. Conversion of any sulfur compounds to hydrogen sulfide and the corresponding hydrocarbons.
4. Saturation of olefinic hydrocarbons, followed by reaction 2 or 3 or both.[115,119,147,160,176,317,337,361]

TABLE 3-2 Sales of Platinum to Consuming Industries, United States, 1968–1973[a]

Industrial Category	Quantity Sold, troy oz					
	1973	1972	1971	1970	1969	1968
Chemical	238,809	225,895	125,112	147,029	175,436	157,677
Petroleum	119,875	98,847	137,396	181,014	58,602	161,050
Glass	72,533	26,970	40,703	34,577	63,350	47,935
Electric	100,607	92,381	51,940	88,146	112,589	117,256
Dental and medical	18,103	30,462	23,097	19,794	22,266	24,903
Jewelry and decorative	20,366	20,655	18,577	30,093	36,151	40,184
Miscellaneous	53,479	50,089	19,859	15,355	47,174	31,150
TOTAL	623,772	545,299	416,684	516,008	515,578	580,155

[a]Data from American Bureau of Metal Statistics[14] and U.S. Bureau of Mines.[213,279,280,472]

TABLE 3-3 Sales of Palladium to Consuming Industries, United States, 1968–1973[a]

Industrial Category	Quantity Sold, troy oz					
	1973	1972	1971	1970	1969	1968
Chemical	239,394	292,710	218,651	186,001	214,508	228,318
Petroleum	3,803	14,499	2,916	15,494	1,337	22,683
Glass	1,439	2,250	237	21,147	3,891	10
Electric	458,652	425,505	431,505	419,089	430,258	329,012
Dental and medical	62,776	94,274	61,594	54,426	52,326	61,636
Jewelry and decorative	21,658	19,375	18,752	17,507	21,837	17,797
Miscellaneous	61,254	27,835	26,451	22,842	34,581	62,023
TOTAL	848,976	876,448	760,106	736,506	758,738	721,479

[a]Data from American Bureau of Metal Statistics[14] and U.S. Bureau of Mines.[213,279,280,472]

In recent years, the trend in this application has been away from the use of platinum alone toward the use of bimetallic catalysts, including mixtures of platinum with rhenium or iridium, and possibly also germanium, indium, or gold. The newer catalysts were reported to permit operation at lower pressures and consequently to result in longer catalyst life and greater octane yield. This application has led to a sharp increase in the demand for iridium.[71,301,338]

Until the introduction of catalytic converters for controlling pollutants emitted from automobile engines, reforming was the major platinum consuming process. Because of its importance, a more comprehensive discussion of catalytic reforming is given in the Appendix.

Exhaust-Gas Control

For some time, ceramic honeycomb materials impregnated with platinum or other platinum metals have been produced and used for exhaust-gas purification. These materials can oxidize a wide array of substances that may be emitted from industrial plants or vehicles. The first applications appear to have been mainly in stationary sources of oxidizable gaseous effluents. Typical of catalysts for this application are a preparation known as Oxycat[211] and another called THT.[2,3] These achieve removal of objectionable volatile substances given off during the curing of a resinous binder, removal of carbon-black fines and other combustible substances given off in the manufacture of carbon black, and oxidation of exhaust gases from petroleum-reforming operations or from the production of phthalic anhydride. On a more limited scale, these materials can be used to remove objectionable fumes from paint-baking ovens, self-cleaning cooking ovens, electric incinerators, etc.[153,440]

The same catalytic devices have been used to remove toxic or offensive fumes from diesel-engine exhaust, so that the engines may be operated in confined areas—for instance, on power fork-lift loaders in factories or warehouses and in the locomotives in mines.[1,400] Similar catalytic devices based on palladium have also been described.

The next logical step has been to apply catalytic units of the foregoing type to reduce hydrocarbon and carbon monoxide emission in automobile exhaust.[4,5] These are based on platinum metals supported on either porous ceramic pebbles or ceramic honeycomb. A suitable combination of these metals has been reported to contain platinum and palladium in a ratio of 5:2,[332] possibly with a small amount of ruthenium or other platinum metal. The use of such catalytic afterburners necessarily involves the use of unleaded fuels, because the lead halides emitted from gasoline that contains lead tetraethyl rapidly poison the catalyst bed. An expanded description of this application is given in Chapter 8. Honeycomb catalyst units of the same type have been applied to the reduction of oxides of nitrogen in exhaust gas. This application was developed to reduce the visible and objectionable emission of nitrogen dioxide from the absorbing towers of nitric acid plants. The tail gas is blended with a reducing fuel gas, such as hydrogen or ammonia, and passed through the catalytic unit, whereupon the nitrogen dioxide is reduced either to nitrous oxide or to elemental nitrogen, according to the concentration of fuel gas added.[219,398] The same principle has been proposed for inclusion in the exhaust train of automobiles. By this it is hoped to remove the nitric oxide, which has been linked to the formation of photochemical smog.[4,404] The catalyst for this purpose generally contains some ruthenium. The technology applying to this particular device does not appear to have reached the point where it can be applied to production automobiles.

Ammonia Oxidation

The manufacture of nitric acid by the Ostwald process entails oxidation of ammonia with air to form nitric oxide. This is accomplished by passing the synthesis gases through gauze beds of an alloy of platinum and 10% rhodium. The operating pressures are about 8 atm (811 kN/m^2).[41,206] There is some small loss of platinum from the catalyst gauze, and this and other changes of the platinum metals have been the object of considerable investigation.[32,49,104,105,182,326,347,393] Although some of the catalyst loss has been attributed to the formation of volatile oxides of the metals at the high operating temperature, some recovery of platinum has been achieved by the use of gold–palladium catchment gauze supported on stainless-steel mesh below the converter catalyst bed.[194,207]

Sulfur Dioxide Oxidation

Although not a current use of platinum, sulfur dioxide oxidation is included here as a reminder that the burning of fuels containing sulfur followed by passage of the combustion products over a platinum catalyst is quite capable of producing sulfuric acid. In fact, the use of platinum catalysts for this reaction resulted in the first patent ever issued in the field of catalysis (cited in Robertson[368]). When used industrially for the production of sulfur trioxide, the early catalysts took the form of either platinized asbestos or finely dispersed platinum on magnesium sulfate. The oxidation of sulfur dioxide is exothermic, so the yield of product decreases as the operating temperature increases. The optimal range of temperature to maintain rapid oxidation and a good yield is about 400–450° C at atmospheric pressure.[316]

Hydrogen Cyanide Manufacture

The Andrussow process for the manufacture of hydrogen cyanide involves the passage of a mixture of air, ammonia, and methane through gauze of an alloy of platinum and 10% rhodium.[329,334] The essential reaction is

$$CH_4 + NH_3 + 3/2\, O_2 \rightarrow HCN + 3H_2O$$

and is exothermic. The yield is approximately two-thirds, on the basis of the consumption of either methane or ammonia. In a newer process, which achieves higher conversion efficiency,[133] only methane and ammonia are used:

$$CH_4 + NH_3 \rightarrow HCN + 3H_2.$$

This exothermic reaction is carried out in alumina reactor tubes coated inside with platinum catalyst.

Hydrogenation

The platinum metals and complex compounds of them have found extensive application as hydrogenation catalysts in organic chemistry. Palladium is the best-known agent for this purpose, but every metal in the group has been found to have advantages for special purposes. Palladium may be introduced as powdered monoxide and reduced *in situ* by the hydrogen, which produces the metal in finely divided form. Alternatively, and more commonly, such inert substances as calcium carbonate, magnesium oxide, activated carbon, and silica gel can be impregnated with soluble palladium compounds. These are allowed to dry and then reduced to yield a supported catalyst. The metal concentration on the support amounts to a few percent at most, except in some special applications. Countless instances of the use of palladium as a catalyst for hydrogenation appear in the literature; a good summary is in a monograph edited by Wise.[385]

Major summaries of this subject have been presented by Rylander,[381] Augustine,[27] and Bond,[51] and shorter summaries by Bond[53] and

Uses

Wells.[471] The use of osmium and ruthenium as hydrogenation catalysts has been described by Bond and Webb[54] and Webb;[466] the characteristics of ruthenium–platinum oxide catalysts for the same purpose have been outlined by Bond and Webster.[55,56] Homogeneous hydrogenation by complexes has been summarized by Rylander.[383]

Dehydrogenation

A wide range of dehydrogenation or oxidation reactions in organic chemistry have been performed with the aid of platinum-metal catalysts. For the most part, these reactions require higher temperatures than hydrogenation to shift the equilibrium toward formation of the desired product. Under these conditions, there is greater risk of side reactions. When a hydrogen-acceptor (such as oxygen or nitrobenzene) is present, the desired reaction may be achieved under more moderate conditions, e.g., at lower temperatures. Palladium appears to be the platinum-group metal most favored for these reactions and commonly is supported on carbon black at 5–30 wt % metal.

The reactions for which this catalytic application is most successful are aromatizations, i.e., conversion of fully or partially saturated molecules into aromatic systems.[382,385] Another kind of catalytic oxidation is the selective oxidation of functional groups, e.g., oxidation of a primary alcohol to an aldehyde; such selective oxidation reactions require the presence of oxygen as a hydrogen-acceptor.

Coordination Complexes of the Platinum-Group Metals as Catalysts

Over a period of not much more than 10 years, there has been extensive development of platinum-metal catalysts of a new type. These are organometallic coordination complexes for the most part, and they are homogeneous catalysts used in a dissolved state. Brief general accounts of these new developments have been presented by Bond[52] and Cleare.[98] Hydrogenation aided by homogeneous platinum-metal catalysts was referred to previously.[383] A recent monograph on platinum and palladium devoted considerable space to a discussion of this newer aspect of their catalytic use.[186(pp. 386–395,443–448)] Many examples appear throughout the monograph by Rylander.[383]

One example of the sort of process made possible in this way is the conversion of methanol to acetic acid with 99% selectivity through insertion of carbon monoxide at pressures as low as 1 atm (101 kN/m^2).[378] The catalyst is a rhodium halide complex with an iodide promoter.

There are some problems in connection with recovery of the expensive catalytic materials when they are used in soluble forms, and some attempts have been made to synthesize "homogeneous" catalysts with polymeric or macromolecular ligands,[267,274] the idea being to create the catalytic center characteristic of a homogeneous agent in a polymerized and therefore insoluble matrix, which is thus capable of ready separation.

Other Industrial or Commercial Catalytic Applications

This section groups a number of other applications of platinum-group metals or their compounds as catalytic agents.

A comparatively new synthesis of hydrogen peroxide is based on the autoxidation of 2-ethylanthraquinol; the quinol is converted to the corresponding quinone and hydrogen peroxide. The latter is separated into water by countercurrent extraction, and the quinone is reduced again to the quinol with a supported palladium catalyst.[24]

The oxidation of ethylene to acetaldehyde with the aid of a palladium chloride, $PdCl_2$, catalyst has become known as the Wacker process.[413,414] The palladium chloride is in aqueous solution with added copper(II) chloride; palladium is reduced to the metal by ethylene, but its reoxidation to the starting compound by a stream of air is aided by the copper chloride. Extension of this principle to the production of acetone and methylethylketone by the oxidation of propylene and butylene, respectively, has also been demonstrated.[103]

Osmium and ruthenium tetroxides have been used as oxidative catalysts in organic chemistry in a variety of ways.[169,170(pp. 66,147),384] One particular reaction for which osmium tetroxide serves as a good catalyst is the hydroxylation of double bonds whereby olefinic compounds are converted to *vic*-glycols with *cis* configuration.[170(pp. 66, 147),384]

New ruthenium catalysts of exceptional activity have been developed for the synthesis of high-molecular-weight polymethylenes from carbon monoxide and hydrogen under pressure. In both two preparations described, the catalyst was metallic ruthenium reduced from an oxide and suspended in nonane.[106]

The production of heavy water involves hydrogen–deuterium exchange on a catalytic surface. A recent account described a dual-temperature exchange between water and hydrogen sulfide; the catalyst consisted of finely divided platinum supported on carbon.[388] In other processes described, the exchange was between gaseous hydrogen and water,[188] with finely divided platinum either in supported form or suspended in the water.

Low-Temperature Catalytic Heaters[261]

Hydrocarbons and light alcohols mixed with air and passed as a vapor over a platinum surface will react, with emission of the heat of combustion. The principle has been used occasionally for such devices as cigarette-lighters. A recent application has been in portable heaters that burn light hydrocarbons. The fuel and air react at an impregnated catalyst bed, producing temperatures of around 400° C. A wide range of recreational, agricultural, and industrial applications of such heaters have been suggested, and a number of manufacturers are producing units based on this principle.

ELECTRIC USES

Contacts for Relays and Switchgear

A large part of the palladium used by industry and utilities is for the production of electric contacts. Telephone switchgear for dialing systems involves millions of small contacts that are expected to provide dependable service for long periods. Palladium, either gold-coated or alloyed with copper or silver, has been adopted in many countries for these switching contacts. The selection of this metal or its alloys is based on good arc-quenching characteristics and minimal sticking, welding, or material transfer—owing to freedom from corrosion or tarnish.

Voltage regulators for automobiles have been improved by the introduction of platinum–iridium, platinum–gold–silver, and palladium–silver–nickel contacts, all of which permit higher currents and extended life. Directional signals incorporate silver and platinum contacts for a variety of reasons, including hardness and good conductivity.

The use of platinum-group metals in electric contacts has been the subject of numerous articles in *Platinum Metals Review*.[39,83,84,167,217,359,386,436,461,463]

Resistors and Capacitors

Windings in traditional wire-wound potentiometers and precision resistors are made from an alloy of palladium and 40% silver, because of the low temperature coefficient of resistance of this composition. Other alloys incorporating the platinum-group metals are also used where other characteristics, such as mechanical strength and higher resistivity, are important.[482]

Many resistors are made by firing solutions containing platinum-metal complexes on glass, quartz, or mica in the form of rods, tubes, plates, or fibers. A wide range of resistances and temperature coefficients of resistance can thereby be obtained. Printed circuits as components of motor controllers, computers, scientific instruments, etc., depend to a considerable extent on plating solutions that contain platinum-group metals, as well as gold and silver.[446,447]

Similar plating and firing techniques have been applied to the fabrication of ceramic capacitors in a multilayer construction.[240,434]

Electrochemical Electrodes

Apart from the familiar use of platinum electrodes in the analytic-chemistry laboratory, platinum and platinum-clad electrodes find many applications in preparative chemistry. In particular, anodes of this material resist oxidation themselves; because there is a high overpotential for the formation of oxygen if they are shiny, many oxidative reactions can be carried out at such a working electrode. For instance, the preparation of hydrogen peroxide by the anodic oxidation of sulfuric acid was widely used, and, although superseded largely by oxidation of 2-ethylanthraquinol,[24] the original electrolytic process has been improved[335] and might be reactivated. A number of electrochemical oxidations forming parts of organic syntheses have been described,[12] and the removal of unwanted chlorides from nitric acid has been achieved electrolytically at platinum anodes.[480]

Although the traditional material for fabrication of electrodes is a combination of platinum and 10% iridium, a number of applications have taken advantage of platinum-clad anodes. For instance, a process has been described for production of sodium hypochlorite from seawater;[446,447] under other conditions, chlorates may be produced at this electrode.[331] Impressed current protection against corrosion will be referred to later, but the same sort of clad electrode finds application for this purpose.

Spark Electrodes

The electrodes in spark plugs for aircraft engines are often made with a platinum alloy. These must operate under extremely corrosive conditions and suffer the risk of early disintegration through the deposition

Uses

of lead from the fuel at the grain boundaries. Alloys with 5 or 10% ruthenium, with palladium and ruthenium together, or quite often with 4% tungsten are used for this purpose.[36] Much-improved performance for the same purpose has been found with pure iridium, although problems of fabricating articles with iridium have delayed its use.[407]

Grids for Power Tubes and Radar Tubes

The grid structures in large thermionic tubes for use in radio transmitters, modulators, and such industrial applications as induction heating are fabricated from platinum-clad molybdenum (or sometimes tungsten) wire. The characteristics of such material include mechanical strength and a high electron work function.[420,446,447]

Fuel Cells

The requirement to generate electricity for satellites and manned space vehicles has stimulated a great deal of research and development of fuel cells. Because these, like conventional chemical cells, are based on converting the energy of a chemical process into electricity, it is not surprising that catalysis plays an important role. It is clear from the literature that platinum-metal electrode systems predominate in the design of fuel cells because of their permanence and their unique catalytic capabilities.[151,306]

Production of Hydrogen

Very recently, the complex tris(2,2′-bipyridine)ruthenium(II)$^{2+}$ has been shown to catalyze water-splitting reactions driven by sunlight.[288] The process works by one of the most efficient mechanisms for converting light energy into chemical energy—photoinduced electron transfer. A key feature of this catalyst is its insolubility in water, which is accomplished by preparing surfactant analogues of bipyridine–ruthenium complex, which can be deposited as monolayer films on glass slides and which will reduce water when subject to ultraviolet irradiation.

HIGH-TEMPERATURE USES

Platinum-Resistance Thermometers

Part of the range of the International Temperature Scale is defined in terms of the platinum-resistance thermometer. This lies between the boiling point of oxygen (−182.97° C) and the melting point of antimony 630.5° C). For this purpose, extremely pure platinum is required with the ratio of its resistances at 100° C and 0° C to be not less than 1.3910:1. A brief review of the construction and applications of the platinum-resistance thermometer has been presented elsewhere,[345] and a more comprehensive monograph has been issued by the National Bureau of Standards.[362]

Thermocouples

For good-quality thermocouples, especially for high-temperature measurement, platinum and platinum–rhodium elements are used. The positive element is the alloy, whose composition must be carefully controlled if existing thermal–electromotive-force data are to be used. Generally, these contain 10 or 13% rhodium. There has recently been a new international agreement on reference tables for platinum-metal thermocouples.[351] For work at higher temperatures, thermocouple elements denoted "five-twenty" or "six-thirty" (referring to the percentage of rhodium in each arm) have demonstrated superior performance. For lower temperatures, thermocouple elements incorporating a palladium–gold alloy (Pallador) have been proposed.[43,309] Thermocouples of these types are used extensively in steel production, nonferrous metallurgy, glass manufacture, the space program, and indeed almost anywhere when accurate information regarding high temperatures is required.

High-Temperature-Furnace Windings

The heating elements in many furnaces designed for use at high temperatures and often under oxidizing or corrosive conditions commonly have resistor wires made of platinum metal. Most are of an alloy of platinum with 10 or 20% rhodium,[174,346] but some trials have been made with an alloy with 40% rhodium,[82] which compares favorably with rhodium itself for this purpose.

Laboratory Ware for High Temperatures

Crucibles, combustion boats, tips of tongs, and other items of laboratory equipment to be used for ignition or other high-temperature work when resistance to chemical attack is important are generally fabricated from an alloy of platinum and 10% rhodium. However, for some purposes, a "nonwetting" platinum alloy may be prepared; this is not "wetted" by molten glasses or borate fluxes. Another recent development has been the introduction of a dispersion-strengthened platinum known as ZGS (zirconia-grain stabilized) platinum,[399] which is stronger and offers more creep resistance than the customary platinum–rhodium alloys.

The preparation of various substances in the form of single crystals for use in lasers, optical modulators, and other devices has increased in importance in the last dozen years. Such crystals may be grown from their components within a flux by slow cooling or by the slow pulling of the crystal from a melt (the Czochrolski technique).[446,447] The crucibles used for either method are made of platinum or iridium, according to the temperature that they are required to withstand. These crucibles can be used at temperatures up to 1350° C for platinum and 2000° C for iridium[101] without contamination of the molten contents.

High-Temperature Strain Gauges

Platinum-alloy resistance wires (tungsten with platinum or chromium with palladium) have been used in strain gauges for high-temperature applications.[47,195]

Flame-Retardants

In apparent conflict with catalytic oxidation activity of platinum, some platinum compounds have been shown to be capable of acting as flame-retardants when included in silicone rubber at very low concentrations.[187,302,406] However, because of the high cost of these materials, it is highly unlikely that the compounds will become widely used for such purposes.

USES BASED ON CORROSION RESISTANCE

There is a wide range of applications related to the corrosion resistance of the platinum-group metals, some based on the intrinsic nobility of the platinum metals themselves, some based on their ability to impart corrosion resistance to or protect other metals, and others based on protection through applied electromotive force, by which a metal to be protected is made cathodic or anodic.

A recent account of the fabrication of standard kilogram weights[417] exemplified the use of these metals because of their freedom from corrosion or tarnish. Sintered platinum-alloy pads for the filtration of highly corrosive fluids are prepared by pressing tiny metallic spheres together and heating them at about 1000° C.[446,447] Platinum-lined furnaces for conducting such chemical reactions as the fluorination of uranium compounds[429] or plutonium oxide[336] with hydrofluoric acid have been described.

The addition of comparatively small amounts of platinum metals has been shown to increase considerably the corrosion resistance of various base metals or alloys. For instance, stainless steel,[202,226] chromium,[201] and titanium[23,111,203,415,416] have been so modified, the first two by the addition of platinum and the last by the addition of palladium. The corrosion resistance of titanium has also been increased by a thin layer of electrodeposited platinum[110,333] or by a thin coating of palladium.[387]

The platinum metals, especially the alloy of platinum and 10% rhodium, are applied in a great variety of ways in the manufacture of glass. Many vessels and furnaces for handling molten glass are fabricated with platinum-clad linings, molybdenum being a preferred support metal. The largest consumption of platinum is for bushings that contain hundreds of small holes through which molten glass is drawn or blown to make glass fiber. Here, corrosion resistance, great strength, and resistance to wear are important. In excellent accounts of the use of the platinum metals in the glass industry,[342-344] it was stressed that the important criterion for their selection is the protection of the purity of the molten glasses.

Finally, an important use for platinum or palladium is an electrolytic method for protection against corrosion. As introduced, this was to protect ships' hulls, propellers, and rudders against corrosion by seawater. These parts of the ship were treated as one electrode of a cell, the other being a number of platinum or platinum-clad electrodes normally on but insulated from the hull; across these electrodes, an electromotive force was impressed such that the hull and other parts were at a potential negative to the other electrodes. This is known as cathodic protection,[446,447] and its use has been extended to steel piers, bulkheads, retaining walls, and pipelines. Base-metal anodes, if used in this way, are corroded away and may be regarded as expendable, but platinum (or titanium coated with platinum or palladium) remains essentially unattacked. The same principle has been used to protect against corrosion in chemical plants, e.g., in the case of paper-making machinery.[13] The principle has been used with reversed polarity to protect steel by anodic current (this is similar to passivation);[198] the counterelectrode is a platinum-clad cathode.

SPINNERETS AND BUSHINGS

As mentioned before, bushings for the production of fiberglass are made from platinum–rhodium alloy.[342-344] Another use of this alloy in the glass industry is in bubbler tubes to be inserted into molten glass. The practice of forcing small gas bubbles through the melt results in agitation and better heat utilization. The bubblers must be able to stand up under severe conditions and are normally made of the alloy of platinum and 10% rhodium welded to a nickel tube.[341]

Platinum alloys are also used to make spinning jets for the production of viscose rayon. These are made with the alloy of platinum and 10% rhodium or, perhaps preferably, with an alloy of gold and 30 or 40% platinum.[191]

FERROMAGNETIC ALLOYS

The alloy containing about 50 atomic percent each of cobalt and platinum is said to make available a more powerful permanent magnet than any other material known. It is also said to possess the advantages (over many other, newer permanent magnetic materials) that it is malleable and ductile before hardening and thereby easily fabricated. On the basis of this material, numerous devices requiring small magnets have appeared, such as hearing aids, electric watches, phonograph pickup cartridges, and miniature relays.[146,277]

HARDENING AGENTS

From what has been said in this chapter and elsewhere in this report, it is clear that, for most applications of massive platinum or palladium (in contrast, for example, with finely divided catalytic material, which is often supported on something else), these metals must be hardened. Traditionally, rhodium and iridium have been the hardening agents for this purpose, the former being particularly suited to applications in which the alloy will be used at high temperatures. The reason for restricting the use of iridium as a hardening agent to alloys intended for cooler uses is related to the measurable losses of iridium when it is heated in air or oxygen (see Chapter 4). In recent years, there has been an increase in the use of ruthenium for the hardening of both platinum and palladium, dictated in part by price considerations.

MEDICAL AND DENTAL USES

The primary medical use of the platinum-group metals in humans today is in cancer chemotherapy, although palladium chloride and palladium hydroxide have been used to treat tuberculosis and obesity, respectively.

In 1965, Rosenberg and co-workers described the bacterial effects of various complexes of platinum.[376] Further studies showed that neutral complexes of platinum, such as *cis*-dichlorodiammineplatinum(II) and a number of congeners (Figure 3–1), inhibited cell division, but not cell growth, in bacteria and that this led to filamentous growth. In 1969, they tested these platinum complexes for antitumor activity against the solid sarcoma-180 tumor in ICR mice and found that about 8 mg/kg of body weight (which is a pharmacologically acceptable dosage) produced complete inhibition of the tumor. The National Cancer Institute

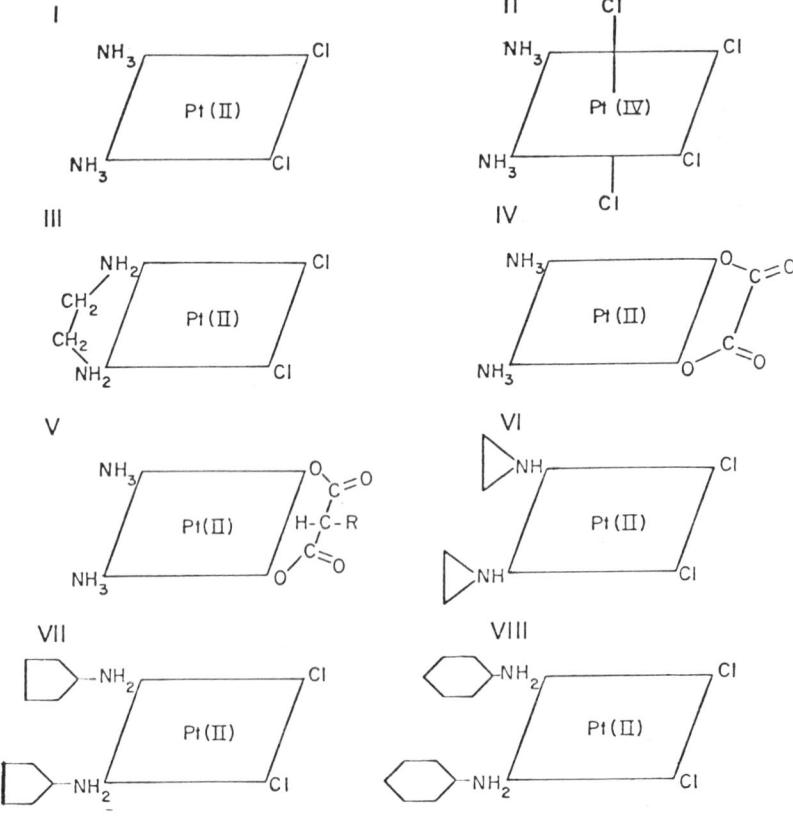

FIGURE 3–1 Structural formulas of some representative antitumor complexes of platinum. I, *cis*-dichlorodiammineplatinum(II); II, *cis*-tetrachlorodiammineplatinum(IV); III, dichloroethylenediammineplatinum(II); IV, oxalatodiammineplatinum(II); V, substituted (R) malonatodiammineplatinum(II); VI, *cis*-dichlorobis(ethyleneimine)platinum(II); VII, *cis*-dichlorobis(cyclopentylamine)platinum(II); VIII, *cis*-dichlorobis(cyclohexylamine)platinum(II).

Uses

(NCI) later found that the same complexes were active against the L1210 tumor in BDF$_1$ mice, confirming their antitumor activity.[377] These complexes—particularly cis-dichlorodiammineplatinum(II), which had been chosen by the NCI for clinical trials—have been extensively tested in many laboratories against a number of model tumor systems (Table 3–4). Platinum drugs appear to be active against a broad spectrum of transplantable tumors, virally induced tumors, and chemically induced tumors and can cause advanced tumors in some systems to regress.

The active complexes have the following characteristics in common: they are neutral coordination complexes of platinum(II) or platinum(IV); they exchange only some of their ligands quickly; two cis monodentate (or one bidentate) leaving groups are required; the corre-

TABLE 3–4 Best Results of Antitumor Activity of cis-Dichlorodiammineplatinum(II) in Animal Systems[a]

Tumor	Host	Best Results[b]
Sarcoma-180 solid	Swiss white mice	T/C = 2–10%
Sarcoma-180 solid (advanced)	Swiss white mice	100% cures
Sarcoma-180 ascites	Swiss white mice	100% cures
Leukemia L1210	BDF$_1$ mice	ILS = 379%; 4/10 cures
Primary Lewis lung carcinoma	BDF$_1$ mice	100% inhibition
Ehrlich ascites	BALB/c mice	ILS = 300%
Walker 256 carcinosarcoma (advanced)	Fisher 344 rats	100% cures; TI > 50
Dunning leukemia (advanced)	Fisher 344 rats	100% cures
P388 lymphocytic leukemia	BDF$_1$ mice	ILS = 533%; 6/10 cures
Reticulum cell sarcoma	C+ mice	ILS = 141%
B-16 melanocarcinoma	BDF$_1$ mice	ILS = 297%; 8/10 cures
ADJ/PC6	BALB/c mice	100% cures; TI = 8
AK leukemia (lymphoma)	AKR/LW mice	ILS = 225%; 3/10 cures
Ependymoblastoma	C57BL6 mice	ILS = 141%; 1/6 cure
Rous sarcoma (advanced)	15-I chickens	65% cures
DMBA-induced mammary carcinoma	Sprague–Dawley rats	77% total regressions 3/9 free of all tumors
ICI 42,464-induced myeloid and lymphatic leukemia	Alderly Park rats	ILS = 400%

[a] Modified from Rosenberg.[373] The data are insufficient to permit useful averages to be presented, with two exceptions: L1210 and B-16. These data represent optimal drug dosages and optimal conditions.

[b] $T/C = \dfrac{\text{tumor mass in treated animals}}{\text{tumor mass in control animals}} \times 100$.

ILS = % increase in life span of treated over control animals.
TI = therapeutic index (LD$_{50}$/ED$_{90}$) × 100; ED$_{90}$ = effective dose to inhibit tumors by 90%.

sponding *trans* isomers of the monodentate groups are inactive; the rate of exchange of the leaving groups should be neither too low nor too high and should fall into a restricted "window of lability" centered roughly on that of the chlorides; and the ligands *trans* to the leaving groups are preferentially strongly bonded, relatively inert amine systems.

The antitumor effect is stereospecific, in that, wherever *cis* complexes have been found to be active, corresponding *trans* complexes have been inactive. The excretion profiles and tissue distribution are approximately the same for both isomers. Instead of differences in the availability of the isomers at specific sites, it appears to be the stereoselectivity of the biochemical reaction in the cell that leads to antitumor activity.

Injected platinum drugs are rapidly excreted, primarily in the urine; the half-life for 80% of a dose is approximately 1.5 h in animals and less than 1 h in humans. The remaining 20% is excreted over a period of weeks. Of the rapidly excreted portion, 95% is unchanged, in the case of both *cis*-dichlorodiammineplatinum(II) and *cis*-malonatodiammineplatinum(II) (the only two tested so far); the other 5% appears to be protein-bound.

The platinum complexes are selectively taken up by the filtering and excretory organs of the body, primarily the kidneys, liver, spleen, and thymus. No selective uptake of platinum complexes in tumor tissue has been shown. The high uptake in the kidneys leads to damage to the proximal convoluted tubules and causes nephrosis in mice and rats. Other forms of toxicosis in these animals are denudation of the intestinal epithelium (leading to the dose-limiting gastrointestinal damage), bone marrow depression, and hypotrophy of the spleen and thymus. No histochemically or physiologically detectable liver damage has been reported in animals or man. Additional forms of toxicosis in man are ototoxicosis (caused by destruction of the cochlear hair cells), nausea and vomiting (due to a central nervous system reaction), and transient anemia. No gastrointestinal toxicosis has been reported in man, and the kidney toxicosis is dose-limiting.

Tissue-culture and *in vivo* biochemical studies have shown that the platinum drugs produce severe and persistent inhibition of DNA synthesis with little or no inhibition of RNA and protein synthesis, at dosages equivalent to therapeutic dosages. The degree of inhibition of DNA synthesis depends on dosage. The synthesis of DNA precursors and their transport through the plasma membrane are not inhibited. DNA polymerase activity is not inhibited. These results are consistent with the generally accepted working hypothesis that the anticancer activity arises from a direct reaction of the platinum drugs with DNA.[373]

Uses

cis-Dichlorodiammineplatinum(II) reacts with DNA *in vitro* in numerous ways. It forms interstrand and intrastrand cross-links. It reacts monofunctionally and bifunctionally with active sites on the bases. It does not appear to form stable products with the phosphates or sugars of the nucleic acids. It is not an intercalating agent. The platinum is mainly localized in regions of the DNA that are rich in guanosine and cytosine. Some of these reactions are reversible or repairable. It is not yet known which type of reaction is significant for anticancer activity.

Gottlieb and Drewinko[165] have reviewed the results of phase I clinical trials of *cis*-dichlorodiammineplatinum(II) in terminally ill cancer patients. The results are compiled in Table 3–5. The overall rate of responses in those trials was generally low, but some types of tumors responded more readily, as listed in Table 3–6.

More recently, the drug has been tested in combination with other drugs. Woodman, Venditti, and co-workers at the NCI showed in 1973[488] that the platinum drug is additive, and in some cases synergistic, with other anticancer agents in its activity against various animal tumors. These results led to the chemotherapeutic use of *cis*-dichlorodiammineplatinum(II) in combination with a wide variety of known anticancer agents in human patients. A second modification in the drug use occurred in 1975, when Cvitkovic and co-workers at the Sloan Kettering Institute developed a simple pharmacologic treatment that largely ameliorated the kidney toxicity of the drug.[116] They hy-

TABLE 3–5 Results of Phase I Clinical Studies with *cis*-Dichlorodiammineplatinum(II) in Terminally Ill Cancer Patients[a]

Investigators	Institution	No. Patients	Responses[b] No.	%
Wiltshaw and Carr	Royal Marsden Hospital	19	7	37
Higby *et al.*	Roswell Park Memorial Institute	50	17	34
Lippman *et al.*	Memorial Hospital for Cancer and Allied Diseases	21	7	33
Hill *et al.*	Wadley Institutes	63	13	21
Rossof *et al.*	Wilford Hall U.S. Air Force Medical Center	21	3	14
DeConti *et al.*	Yale University	10	1	10
Talley *et al.*	Southwest Oncology Group	57	5	9
Kovach *et al.*	Mayo Clinic	51	2	4

[a] Data from Gottlieb and Drewinko.[165]
[b] Greater than 50% reduction in tumor mass.

TABLE 3-6 Responses of Specific Cancers to
cis-Dichlorodiammineplatinum(II)[a]

Diagnosis	No. Patients	No. Responses			Total Response Rate, %
		Complete[b]	Partial[c]	Improvements[d]	
Testicular carcinoma	16	7	3	3	81
Lymphoma	16	2	7	1	63
Squamous cell carcinoma of the head and neck	17	0	1	6	41
Ovarian carcinoma	20	0	5	3	40

[a] From studies shown in Table 3-5.
[b] Other complete responses: bladder carcinoma (2); thyroid carcinoma (1).
[c] Not complete, but ≥50% tumor reduction. Other partial responses: breast carcinoma (2); acute myelogenous leukemia, endometrial carcinoma, renal carcinoma, thymoma, neuroblastoma, lung adenocarcinoma, unknown undifferentiated primary (1 each).
[d] ≤50% tumor reduction, significant subjective improvement, or mixed response. Other improvements: colon (3); multiple myeloma, breast carcinoma, acute myelogenous leukemia, lung carcinoma, prostatic carcinoma, unknown undifferentiated primary, undifferentiated sarcoma (1 each).

drated the patients with D-mannitol, an osmotic diuretic, before treating them; this apparently protects the kidneys, but does not increase the percentage excretion of the drug. Other side effects are now more prominent. Myelosuppression now appears to be a dose-limiting factor. With this hydration treatment, the same workers have been able to increase the dosage safely by a factor of 3. With the high dosage in combination with other drugs, they have achieved a 95% remission rate in patients with testicular cancer. The duration of the remission is at least 14 months, and the relapse rate is very low. The best previous therapy produced a median life extension of 6 months. Thus, the addition of the platinum drug to the prior combination therapy has produced significant increases in the number, extent, and duration of remissions. The sensitivity of different kinds of testicular cancers to the platinum drug has now been verified by Einhorn et al.[132] at the Indiana University Medical Center and by Merrin[273] at the Roswell Park Memorial Institute.

Wiltshaw and Kroner[484] have shown that ovarian carcinoma is also responsive to the platinum drug. Bruckner and co-workers[68] at the Mt. Sinai School of Medicine have achieved a 70% response rate in ovarian carcinoma patients with a combination of the platinum drug (at low dosages) and adriamycin. Some recent clinical results with combination and high-dose cis-dichlorodiammineplatinum(II) therapy, collected in Table 3-7, show the potential value of this drug.

It is possible that a patient who was not sensitive to the platinum drug initially may become hypersensitive after a series of injections. This was apparently observed in at least one case. However, little is known about the sensitivity reaction, and anaphylactoid reactions can be attributed to impurities in the drug. Since the NCI has become more conscious of the need for high purity of the drug, these reactions have not appeared. The neutral species (the active drugs) do not seem to generate allergic reactions.

The clinical results seen thus far indicate that the search for less toxic but more effective analogues should be continued. In addition, the recent results of combination therapy suggest that this may be more effective than the use of platinum compounds as single agents (see Table 3-7). A number of reviews and symposia[107,372,373] have reported in detail the animal antitumor studies, animal toxicology, analogue development, mechanisms of action, and clinical trials of the platinum coordination complexes.

Various amounts of silver, copper, zinc, platinum, and palladium are added to gold by manufacturers to change such properties as hardness, strength, color, and cost. Gold is alloyed with other metals to improve its physical properties; the products are the strongest and most versatile restorative materials used in dentistry. Generally, gold alloys are based on a three-part composition—70% gold and the remainder copper and silver. To ensure hardness and to make heat treatment possible, the proportion of copper is increased. To offset the lowering of the melting-point range caused by the increase in copper content, platinum or palladium is added; this addition may also improve the results of the heat treatment. An addition of 10% platinum to a simple ternary alloy increases its strength by 35%. However, owing to the high cost of platinum, its place in dental alloys has been taken to some extent by palladium. Wrought-gold alloys may contain up to 10% palladium. In the so-called white golds, 30-40% of the gold is replaced by palladium; this lowers their resistance to tarnish, as well as lowering the cost.[21,380]

MISCELLANEOUS USES

Separation of Pure Hydrogen

Palladium has the remarkable property (described more fully in Chapter 4) of being able to absorb or desorb hydrogen gas. Advantage is taken of this property in the development of units to separate hydrogen

TABLE 3-7 Preliminary Clinical Trials of cis-Dichlorodiammineplatinum(II)

Tumor Type	Drug Therapy	No. Evaluable Patients	Complete Remissions, % (Duration)	Partial Remissions (>50%, <100%), % (Duration)	Total Responses, %
Testicular cancers (metastatic)[a]	cis-Dichlorodiammine-platinum(II) + vinblastine + bleomycin	39	85 (3+ to 24+ mo.)	15 (3+ to 24+ mo.)	100
Epidermoid carcinoma of head and neck (far advanced)[b]	High-dose cis-dichlorodiammineplatinum(II)	26	8 (2+, 6+ mo.)	23 (1,2,3,4,5+,6+ mo.) 38 (measurable responses)	69
Ovarian carcinoma (advanced; prior therapy failed)[c]	cis-Dichlorodiammine-platinum(II) + adriamycin	18	33	33	67 (89% survival in remission)
Bladder cancer[d]	cis-Dichlorodiammine-platinum(II)	24	0	33 17 (measurable responses)	50

[a] Data from Einhorn and Furnas.[131]
[b] Data from Wittes et al.[486]
[c] Data from Bruckner et al.[67]
[d] Data from Yagoda et al.[492]

Uses

from other gases by passing gases that contain hydrogen through diffusion barriers of palladium or a silver–palladium alloy. The hydrogen produced is ultrapure, and the purification units have been developed from laboratory scale to industrial capacity. This technique and its practical applications have been described in *Platinum Metals Review*.[48,102,117,220,348,367,430,449]

Jewelry

The use of platinum for jewelry appears, from statistics on sales to the industry, to have declined over the last few years; such use of palladium has remained about constant. Nevertheless, as Tables 3-2 and 3-3 show, the demand is still substantial. Platinum is still used for small articles—such as rings and settings for jewels—in the form of a hardened alloy, usually with 5-10% iridium. Because of its lower cost, palladium has some appeal, and a limited range of luxury goods—such as jewelry, cigarette cases, and the like—are produced from this metal, generally alloyed with 4.5% ruthenium or 4% ruthenium and 1% rhodium. White gold is an alloy that contains about 20% palladium or nickel.

Reflecting or Ornamental Surfaces

There is a considerable demand for articles plated with rhodium. It has a hard and highly reflecting surface. Although the reflectivity of rhodium is not as great as that of silver, the absence of tarnish offers a great advantage over silver. Many items of silverware are rhodium-plated to protect them from the otherwise inevitable dark tarnish. Other items requiring a good-looking and durable surface—e.g., camera fittings and jewelry—are similarly coated, either electrolytically or by vacuum deposition.

Mirrors and reflectors are often plated with rhodium, especially if the conditions for their use may be corrosive, e.g., in scientific equipment or lighthouse reflectors.[360] Platinum has been applied on and off for nearly a century and a half to produce a silvery luster on ceramic glazes, and an interesting contemporary account was given in a short article by Wynn.[491] In addition to this decorative use, a platinum–gold layer fired onto the inner surface of some mercury vapor lamps will reduce their infrared radiation through multiple internal reflection.[354]

Brazing Alloys

Many brazing alloys contain noble metals. One—containing 20% palladium, 5% manganese, and the remainder silver—has been used to join the thin-walled tubing in the thrust chambers of the F-1 rocket engine used, for instance, on the Apollo spacecraft.[446,447] Another, containing palladium with silver and copper, has been used for the multistage jointing in the manufacture of heavy-duty power tubes, such as magnetrons and klystrons, for high-temperature operation.[410,411] Another alloy for brazing tungsten consisted mainly of platinum with a few percent of boron.

Protective Bursting Disks

For the protection of chemical process equipment, rupture diaphragms or bursting disks (which burst if the pressure inside the equipment exceeds a safe value) are often used. Platinum is the most useful material for the diaphragm itself, because its great malleability permits sheets to be rolled to a uniform and specified thickness and because it undergoes minimal change during operation, owing to the resistance of the metal to corrosive attack.[330] A number of examples of this application have been described.[325]

Gamma Radiography with Iridium-192

It has been shown[227,278] that, for testing castings and welded structures by radiography, iridium-192 offers several attractive characteristics and may be used instead of cobalt-60.

Histologic Stain

Use as a histologic stain is peculiar to osmium tetroxide, which has been so applied for many years. The conditions in the tissue are such as to bring about reduction of osmium to the metal in a dark, insoluble form that is ideal for highlighting the structure.[169]

Fountain-Pen Nibs, Instrument Pivots, etc.

For a long time, osmium alloys were produced that, because of their great hardness, were used for special purposes, e.g., in nibs for fountain pens, in long-life phonograph needles, and in instrument pivots. Their use in such applications has been all but completely superseded by changes in fashion or by introduction of other materials. The use of a ruthenium alloy in the form of tiny balls (0.02–0.04 in., or 0.5–1.0 mm, in diameter) at the tips of ball-point pens has been described.[446,447]

4

Physical and Chemical Properties

PHYSICAL PROPERTIES

A selection of physical properties of the platinum-group metals is given in Table 4–1. The properties cited in the table are related, insofar as possible, to the purest available specimens of metal. It is well known, however, that minute traces of impurities cause appreciable changes in such physical properties of these metals as hardness and electric resistivity. Because the platinum metals have a marked tendency to absorb such gases as hydrogen and oxygen, which have a significant influence on the physical properties of the metals, it is easy to imagine the difficulty in establishing some values. Mechanical properties of these metals, and others, depend on the amount of cold working that has preceded their measurement. The mechanical and tensile properties of osmium and ruthenium are anisotropic, and this is attributed to inequalities in spacing in their hexagonal close-packed structures.

For many years, osmium was described as the densest element in the periodic table, but more recent measurements show that iridium is denser, by about 0.2%. Osmium shows the minimal atomic volume among the transition elements of the third long period. It was also believed for many years that ruthenium exhibited allotropy; a transition temperature of 1035° C was thought to be detectable between two forms. It is now concluded that the allotropy does not exist.[215]

An artificial radioisotope of iridium, iridium-192, may be mentioned briefly because of its practical application as a source in industrial

TABLE 4-1 Some Physical Properties of the Platinum-Group Metals[a]

Property	Platinum	Iridium	Osmium	Palladium	Rhodium	Ruthenium
Atomic number	78	77	76	46	45	44
Atomic weight	195.09	192.22	190.2	106.4	102.9055	101.07
Stable isotopes (% abundance)	192 (0.78)	191 (38.5)	184 (0.018)	102 (0.8)	103 (100)	96 (5.7)
	194 (32.8)	193 (61.5)	186 (1.59)	104 (9.3)		98 (2.2)
	195 (33.7)		187 (1.64)	105 (22.6)		99 (12.8)
	196 (25.4)		188 (13.3)	106 (27.2)		100 (12.7)
	198 (7.23)		189 (16.1)	108 (26.8)		101 (17.0)
			190 (26.4)	110 (13.5)		102 (31.3)
			192 (41.0)			104 (18.3)
Density (at 20° C), g/cm^3	21.45	22.65	22.61	12.02	12.41	12.45
Crystal lattice (closest packed)	Cubic	Cubic	Hexagonal	Cubic	Cubic	Hexagonal
Lattice constants (at 20° C), Å						
a	3.9229	3.8392	2.7340	3.8906	3.8029	2.7056
c/a	—	—	1.5799	—	—	1.5825
Melting point, °C	1768	2443	3050	1552	1960	2310
Thermal conductivity, W/cm-°C	0.73	1.48	0.87	0.76	1.50	1.05
Linear coefficient of thermal expansion × 10^6 (at 20–100° C), per °C	9.1	6.8	6.1	11.1	8.3	9.1
Specific heat (at °C), cal/g-°C	0.03136	0.0307	0.0309	0.0584	0.0589	0.0551
Heat capacity (C_p, at 25° C), cal/mole-°C	6.18	6.00	5.90	6.21	5.98	5.75

Entropy (S, at 25° C), cal/mole-°C	9.95	8.48	7.8	9.06	7.53	6.82
Latent heat of fusion, kcal/mole	4.7	6.3	7.6	4.2	5.15	6.2
Latent heat of evaporation, kcal/mole	135.0	160.0	162.0	84.3	133.1	155.0
Electric resistivity (at 0°C), μohm-cm	9.85	4.71	8.12	9.93	4.33	6.80
Temperature coefficient of resistance (at 0–100° C), per °C	0.0039	0.0043	0.0042	0.0038	0.0046	0.0042
Thermal-neutron cross section, barns	9±1	425±15	15.3±0.7	6.0±1.0	150±5	3.0±0.8
Hardness (annealed), vHN	40–42	200–240	300–670	40–42	100–102	200–350
Tensile strength (annealed), tons/in.²	9	80	—	12.5	50	36
Young's modulus (annealed), tons/in.² × 10⁻⁴	1.2	2.75	4.0	0.85	2.3	3.0
Magnetic susceptibility (χ), cm³/g × 10⁶	0.9712	0.133	0.052	5.231	0.09903	0.427
Work function (φ), eV	5.27	5.40	4.8	4.99	4.90	>4.54
Thermionic function (A), amp/cm²-K	64	170	120	60	100	—

[a]Data from *Platinum Metals Review*,[327] Tugwell,[447] and Goldberg and Hepler.[162]

radiography of ferrous welds. It emits gamma rays with average energy of 0.40 MeV and has a half-life of 72 days.[278]

The numerical values of some properties given in Table 4–1 (e.g., melting point, latent heat of evaporation, and hardness, or mechanical strength) reveal a progressive decrease in coherence or bond strength among atoms in each triad (ruthenium, rhodium, and palladium; osmium, iridium, and platinum) as atomic number increases. Such trends can be correlated with the progressive decrease in the number of electrons available for bonding in the solid state. These will occupy orbitals that have been hybridized from s, p, and d states. The amount of d character that can be contributed to these orbitals is presumed to decrease as electrons become paired in atomic d orbitals or otherwise fail to engage in metallic bonding. The tendency described continues beyond the group VIII metals to the corresponding members of groups IB and IIB.[340]

Radioactive isotopes of ruthenium are produced in the fission of uranium-234, and a few of the characteristics of these species will be mentioned here. There are four known radioisotopes of ruthenium. Ruthenium-97 (half-life, 2.9 days; decay via electron capture and gamma-ray emission) and ruthenium-105 (half-life, 4.4 h; decay via electron capture and moderate beta emission) are formed only by neutron activation. Ruthenium-106 (half-life, 1 yr; decay via low-energy beta emission) is formed only by decay of uranium-235, but its daughter, rhodium-106 (half-life, 30 s), is a high-energy beta-emitter. Ruthenium-106 and rhodium-106 are sometimes used as quality-control monitors to measure the thickness of sheets of plastic and metal produced on a continuous basis. The last isotope, ruthenium-103 (half-life, 40 days; decay via weak beta emission), can be formed either by neutron activation or by uranium-235 fission, where it represents 2.9% of the fission yield. The fission yield of ruthenium-106 is only 0.38%.

Because such small quantities of ruthenium are produced during fission, the only potential health problems that could be associated with this material are due to its radioactivity. Current practice is to retain the ruthenium as a sulfide in a holding tank until its radioactivity has decayed to a safe point.[141,483]

ALLOYING CHARACTERISTICS

The properties of alloys of the platinum metals with each other and with other metals have been extensively studied (see, e.g., Vines and Wise,[459] numerous technical bulletins issued by the International Nickel Company, and of course *Platinum Metals Review*). The hardness of platinum is deliberately raised for many of its uses by the addition of

Physical and Chemical Properties

alloying elements. Of these, nickel, osmium, ruthenium, copper, gold, silver, and iridium all produce considerable increases in hardness, the effect per unit weight of added element decreasing approximately in the order named. Rhodium and palladium produce much less increase in hardness than do the preceding metals. Commercially important alloys of platinum are prepared with copper, gold, iridium, rhodium, and ruthenium. In recent years, alloys with cobalt have become important because of their strong ferromagnetic properties.

In much the same way, palladium forms alloys with advantages over the pure metal for various practical applications.[485] In general, the alloying elements tend to increase the resistivity, hardness, and tensile strength of palladium. Copper, nickel, gold, iridium, rhodium, and ruthenium have been used in this way, and an alloy with silver has been widely used in electric contacts.

Extensive or complete miscibility of one metal with another is generally favored by similarity in crystal structure between the two metals, similarity in atomic radii, and similarity in valence. On the basis of these three criteria, extensive mutual solubility would be expected among the four platinum metals with cubic close-packed structures (see Table 4-2). But limits of solubility, imposed by differences in crystal structure, would be expected for alloys of these four metals with ruthenium or osmium. In fact, platinum and palladium form solid solutions in all proportions with the elements making up groups VIII and IB of the periodic table, except ruthenium and osmium. The complete miscibility of iron with platinum or palladium is evidently associated with stabilization of the gamma modification of iron by small amounts of the second element. Iron alloys containing as little as 5% of palladium or platinum solidify as cubic close-packed crystals.

The formation of alloys by the platinum metals has been discussed in terms of the numbers of electrons available for bond formation in the solid state.[216] As indicated earlier, these become fewer, with consequent lack of cohesion, in passing from ruthenium to palladium or from osmium to platinum. The role of the hardening elements added to palladium or platinum is to increase the pool of d electrons and so augment the strength of the metallic bond. Some of the differences in alloying between platinum and palladium suggest that the latter has the smaller number of valence electrons and lower proportion of d character in its bonding. Such conclusions are consistent with the chemical evidence of preferred bivalence in palladium compounds and of preferred quadrivalence in platinum compounds.

There is an unexpected, and as yet unexplained, feature of alloy formation in platinum and palladium: the occurrence of extensive miscibility gaps in some alloy systems prepared at lower temperatures,

TABLE 4-2 Atomic Radii and Crystal Types of Group VIII and IB Elements

Property	Group VIIIA	Group VIIIB	Group VIIIC	Group IB
	Iron	Cobalt	Nickel	Copper
Atomic radius, Å	1.27	1.25	1.245	1.275
Crystal type[a]	α BCC γ CCP	CCP	CCP	CCP
	Ruthenium	Rhodium	Palladium	Silver
Atomic radius, Å	1.335	1.34	1.375	1.442
Crystal type[a]	HCP	CCP	CCP	CCP
	Osmium	Iridium	Platinum	Gold
Atomic radius, Å	1.35	1.354	1.385	1.439
Crystal type[a]	HCP	CCP	CCP	CCP

[a] BCC = body-centered, cubic; CCP = cubic, closest-packed; HCP = hexagonal, closest-packed.

whereas the same systems prepared at higher temperatures exhibit complete miscibility. Thus, palladium forms a continuous series of solid solutions with rhodium and iridium at high temperatures, but appreciable gaps in miscibility have been found in the same system below 850° and 1500° C, respectively.[353] The precise compositions of the miscibility limits have not yet been established.

The alloying behavior of the other four platinum metals follows for the most part the general principles outlined. Less is known about the characteristics of osmium alloys, owing to difficulties in working with this hard and high-melting-point element. Rhodium displays an unexpected reluctance to form solutions with silver and gold, but otherwise is completely miscible with the other cubic close-packed elements in Table 4-2, including γ-iron. The alloys containing about 50 atomic percent each of iron and rhodium have an unusual magnetic property: they are practically nonmagnetic at room temperature, but suddenly become ferromagnetic when heated to 60° C.

EFFECT OF SUBDIVISION ON CHEMICAL REACTIVITY

Platinum is relatively inert, with respect to chemical attack by oxygen or many acids, and a number of its uses are based on this property.

Physical and Chemical Properties

However, it should be stressed that the chemical reactivity of platinum and the other elements in the group is markedly influenced by the state of subdivision of the metal. Thus, the sponge obtained by igniting ammonium chloroplatinate is more readily attacked than the compact metal. Similarly, platinum dissolved in another metal, such as lead or silver, is much more readily attacked. Platinum black or platinum finely dispersed on a porous bed, such as silica gel, is still more reactive and displays remarkable catalytic properties. Similar comments apply to the other members of the group.

The "nobility" of platinum and its congeners arises from their bonding in the solid state. Thus, the first-stage ionization potential of platinum is 9.0 V, only a trifle higher than that of other transition elements. The standard electrode potential (E°)—for 2e + Pt^{2+} (aq) = Pt (s)—has been estimated to be +1.2 V. An analysis of the energy quantities leading to such a value of E°, in the manner of the Born–Haber cycle, discloses that the significant term is the very high sublimation energy of the crystal. The increase in reactivity exhibited by samples with a high specific surface area, compared with that of bulk metal, can be attributed to an increase in the number of atoms with the higher energy associated with surface sites.

CHEMICAL REACTIONS WITH OXYGEN, HALOGENS, AND ACIDS

The direct oxidation of the platinum-group metals is summarized in Table 4–3. The formation of the volatile osmium tetroxide by finely divided osmium occurs at room temperature and can be detected by its distinctive odor. Although ruthenium also forms a volatile tetroxide, it differs from osmium, in that its tetroxide does not form directly from the elements at moderate temperatures.

Four of the metals other than osmium sustain a detectable weight loss when heated in oxygen at high temperatures. Thus, platinum loses some weight in oxygen at 1000° C. No such loss is observed when it is heated to this temperature in a vacuum or in an inert gas. This observation is attributed to the removal, through either volatilization or decomposition, of a film of platinum dioxide, which probably coats the metal even at room temperature.[85–87,122,260,448] Similar losses in weight occur with rhodium and iridium when they are heated in air or oxygen to temperatures above 1100° C, the effect being greater in the case of iridium. The weight loss is explained by the formation of volatile rhodium dioxide, RhO_2, or iridium trioxide, IrO_3. Ruthenium, when heated in air at 1000° C, sustains the greatest weight loss, believed to

TABLE 4-3   Reaction of Platinum-Group Metals with Pure or Atmospheric Oxygen

Metal	Extent of Oxide Formation	Oxide Formed		Formation Temperature, °C	Decomposition Temperature, °C
Platinum	Negligible	Platinum(IV) oxide Platinum dioxide	} PtO_2	<1000	—
Palladium	Superficial	Palladium(II) oxide Palladium monoxide	} PdO	>350	>870
Rhodium	Superficial	Rhodium(III) oxide Rhodium sesquioxide	} Rh_2O_3	~700	1100
Iridium	Superficial	Iridium(IV) oxide Iridium dioxide	} IrO_2	~700	1140
Osmium	Considerable	Osmium(VIII) oxide Osmium tetroxide	} OsO_4	200	—
Ruthenium	Superficial	Ruthenium(IV) oxide Ruthenium dioxide	} RuO_2	700	—

be due to formation of volatile ruthenium trioxide, RuO_3, with a vapor pressure of 10^{-3} atm (101 N/m²) just above 1100° C.

Palladium has the capacity to absorb hydrogen—as much as 900 times its own volume under standard conditions—over a range of temperatures.[255] The uptake of hydrogen corresponds roughly to the composition Pd_2H, but modern studies appear to have largely ruled out the formation of such a discrete substance. Instead, it is inferred that below 300° C there are two phases, each consisting of a solid solution, whereas above this critical temperature there is only a single solution phase. In each phase, hydrogen atoms are held interstitially in such a way as to involve actual chemical bonding, as deduced from changes in electric conductance and magnetic susceptibility. To a smaller degree, platinum and rhodium exhibit a similar absorption of gaseous hydrogen.

The principal products of direct reaction of the halogens with the heated platinum-group metals are shown in Table 4–4. The temperatures shown are those recommended for good yield, but are not critical to the formation of some product. Even at room temperature, palladium is corroded by moist chlorine or bromine, and a palladium anode is appreciably dissolved during electrolysis of chloride solutions. Likewise, saturated chlorine or bromine water or an alcoholic solution of iodine will corrode metallic ruthenium.

Table 4–5 summarizes the action of acids on the platinum-group metals. Resistance to attack by common acids is shown in increasing

TABLE 4-4 Products of Reactions of Platinum-Group Metals with Fluorine and Chlorine

Metal	Reaction with Fluorine		Reaction with Chlorine	
	Products	Temperature, °C	Products	Temperature, °C
Platinum	Platinum tetrafluoride, PtF_4	500	Platinum dichloride, $PtCl_2$	500
			Platinum trichloride, $PtCl_3$	
			Platinum tetrachloride, $PtCl_4$	
Palladium	Palladium trifluoride, PdF_3	500	Palladium dichloride, $PdCl_2$	300
Rhodium	Rhodium trifluoride, RhF_3	500	Rhodium trichloride, $RhCl_3$	600
	Rhodium pentafluoride, RhF_5	400		600
Iridium	Iridium pentafluoride, IrF_5	360	Iridium trichloride, $IrCl_3$	600
	Iridium hexafluoride, IrF_6	300		
Ruthenium	Ruthenium pentafluoride, RuF_5	300	Ruthenium trichloride, $RuCl_3$	400
Osmium	Osmium hexafluoride, OsF_6	300	Osmium trichloride, $OsCl_3$	<500
			Osmium tetrachloride, $OsCl_4$	>650

TABLE 4-5 Attack of Platinum-Group Metals by Mineral Acids

Metal	Form	Nature of Attack
Palladium	Compact	Attacked by hot concentrated nitric acid and boiling sulfuric acid; dissolved by aqua regia
	Sponge	Dissolved by all the above acids
Platinum	Compact or sponge	Not attacked by single mineral acids; dissolved by aqua regia
Rhodium	Compact	Attacked by boiling sulfuric acid or hydrobromic acid; not dissolved by aqua regia
	Dispersed	Dissolved at least partially by aqua regia
Iridium	Compact	Practically unattacked by hot mineral acids or aqua regia
	Sponge	Dissolved in Carius tube by hot hydrochloric acid plus an oxidizing agent (nitric acid or sodium chlorate)[475]
Ruthenium	Any	Virtually unattacked by hot mineral acids or aqua regia
Osmium	Any	Virtually unattacked by hot mineral acids or aqua regia

order in this table. The action of aqua regia on palladium and platinum yields chloropalladic acid, H_2PdCl_6, and chloroplatinic acid, H_2PtCl_6; however, on evaporation of a solution of the former, the dichloride is the compound recovered. Rhodium can be rendered susceptible to attack by aqua regia if it is dispersed in an alloy that is dissolved by this acid.

In contrast with the acid reactions just given, the platinum-group metals show a variety of responses to alkaline fusions, especially in the presence of oxidizing agents. Thus, sodium peroxide or a mixture of sodium hydroxide and sodium chlorate will bring osmium or ruthenium into soluble forms, usually sodium osmate, Na_2OsO_4, or sodium ruthenate, Na_2RuO_4. These reactions are best carried out with finely divided metal. These metals are also attacked appreciably even by alkaline hypochlorite solutions. Rhodium is also attacked, but to a smaller degree, by such alkaline fusions, as well as by fused alkali cyanides. Platinum and palladium are appreciably corroded under conditions of alkaline fusion or by fused alkali cyanide. The effect of such treatment on iridium is the least among the group of metals, but it is not negligible.

The behavior of platinum toward a number of elements at high temperature is or should be familiar to users of platinum laboratory ware. Carbon, phosphorus, silicon, arsenic, etc., combine or alloy with hot platinum, so care must be taken in heating compounds of these ele-

Physical and Chemical Properties

ments to avoid reducing conditions. Contact with sooty burner flames or unburned gas may lead to embrittlement, owing to formation of a carbide.

Ruthenium and iridium have been shown to resist chemical attack by a number of molten metals when heated in argon atmospheres. Thus, crucibles of these metals, especially iridium, can be used to contain a number of normally very reactive elements at high temperatures.

SELECTED COMPOUNDS

Binary Compounds

PLATINUM

The principal oxidation states of platinum are +2 and +4; of these, the first is the more common. Most platinum compounds are coordination complexes; there is no compelling evidence of the existence of simple aqueous metal ions. The coordination chemistry of platinum is outlined later in this chapter. Some of the simpler binary compounds are described here, with brief mention of some of the related complex ions, e.g., those with the halogens. In these complex ions, bivalent platinum assumes a coordination number of 4 (square planar), and quadrivalent platinum, 6 (octahedral).

Platinic chloride, $PtCl_4$, is a red-brown crystalline solid that can be formed by high-temperature chlorination of platinum, but is more conveniently prepared by decomposing chloroplatinic acid with heat in a stream of hydrogen chloride (165° C) or chlorine (369° C). It is readily soluble in water, alcohol, and acetone.

Chloroplatinic acid, H_2PtCl_6, is formed when platinum is dissolved in aqua regia. It is a dark-red crystalline solid whose aqueous solution is yellow or orange, according to concentration; in solution, it acts as a strong acid. The salts potassium hexachloroplatinate, K_2PtCl_6, and ammonium hexachloroplatinate, $(NH_4)_2PtCl_6$, are sparingly soluble, and the former has been used in the gravimetric analysis of platinum. Many organic amines also form insoluble chloroplatinates that may be used for their characterization.

Platinous chloride, $PtCl_2$, is a brownish-green solid usually prepared by heating platinic chloride in an atmosphere of chlorine (580° C). It is insoluble in water, but dissolves in hydrochloric acid to form a dark brown solution of chloroplatinous acid, H_2PtCl_4. The latter may also be prepared by reduction of chloroplatinous acid with sulfur dioxide.

There are two fluorides of some importance. Platinum hexafluoride, PtF_6, is a dark-red solid (m.p., 61.3° C) with only a narrow range of existence in the liquid state (b.p., 69.1° C); its vapor is brown. It can be prepared by electrically heating platinum wire in fluorine close to a surface cooled by liquid nitrogen. Being thermally unstable, it breaks down by way of an unstable pentafluoride, PtF_5, to PtF_4. The hexafluoride is an extremely powerful oxidizing agent; its place in chemical history is ensured as the substance that first oxidized an inert gas ($Xe + PtF_6 \rightarrow XePtF_6$). The tetrafluoride, PtF_4, may be prepared by treating platinous chloride with fluorine at 200° C; it is a yellow-brown solid, slowly hydrolyzed by water.

Platinic oxide, PtO_2, may be prepared by treating chloroplatinic acid with sodium carbonate; the resulting residue is extracted with acetic acid, and the insoluble remainder, consisting of yellow platinic acid, $H_2Pt(OH)_6$ [or $Pt(OH)_4 \cdot 2H_2O$], is heated below 100° C to yield the black dioxide. When this is heated more strongly, the metal is obtained. The yellow "hydroxide" is known as platinic acid; it is amphoteric, dissolving in either hydrochloric acid or alkali.

Platinous oxide, PtO, may be obtained by carefully heating the black hydroxide formed by the addition of an alkali to a chloroplatinite. This gray oxide is subject to disproportion if heated too strongly, and platinum and platinic oxide are formed. Treatment of the hydroxide with hydrochloric acid again leads to disproportionation, with platinum and chloroplatinic acid being formed.

Platinic sulfide, PtS_2, is obtained as a black precipitate when hot acidified chloroplatinate solutions are treated with hydrogen sulfide. This compound is soluble in alkaline polysulfide solutions, and it is another form in which platinum may be precipitated for gravimetric analysis.

PALLADIUM

In its compounds, palladium most commonly exhibits an oxidation state of +2, although it may less commonly be quadripositive (+4), and in a few instances terpositive (+3). A few univalent complexes of palladium have also been reported.[186,307] Like the other platinum metals, it has a strong disposition to form coordination complexes. Those of bivalent palladium show a coordination number of 4 for this element and a square planar structure.

Palladous oxide, PdO, appears as a black powder when palladium sponge is heated in oxygen. It may also be prepared (for instance, in making palladium catalyst) by fusing palladous chloride, $PdCl_2$, with potassium nitrate at 600° C and then leaching out the water-soluble

Physical and Chemical Properties

residue. The oxide is insoluble in water and boiling acids (including aqua regia). It can easily be reduced by heating in hydrogen, and the metal so produced is an active hydrogenation catalyst. When alkali is added to aqueous palladous salts, a yellow hydrous oxide, $Pd(OH)_2$, is produced. This loses water and turns black when heated to 500° or 600° C. The dissociation pressure of palladous oxide reaches 1 atm (101 kN/m^2) at 875° C.

Palladous chloride may be prepared by direct union of the elements at 500° C; above 600° C, it dissociates to the elements. It is a red deliquescent solid; from its aqueous solution or from solutions of palladium dissolved in aqua regia, crystals of palladous chloride dihydrate, $PdCl_2 \cdot 2H_2O$, may be obtained. Evidence of the existence of the corresponding acid—chloropalladous acid, H_2PdCl_4—is open to question, but salts derived from it, such as K_2PdCl_4, are obtained by adding the stoichiometric amount of the appropriate metal chloride to aqueous palladous chloride and evaporating to dryness. The corresponding bromide and iodide are dark solids, insoluble in water, but dissolved by an excess of the halide ion as complex ions, PdX_4^{2-}.

Fluorine reacts with metallic palladium or with palladous chloride to form palladium trifluoride, PdF_3, a black solid. This is an active oxidizing agent. On reduction, it yields palladous fluoride, PdF_2, generally contaminated with palladium. Pure PdF_2 may be produced by treating palladium trifluoride with selenium tetrafluoride; it is a violet crystalline solid, completely hydrolyzed by water.

Palladium forms a number of compounds with sulfur (and also with selenium and tellurium). For example, the following sulfides have been deduced from the phase diagram and, in part, characterized by x-ray measurements: Pd_4S, $Pd_{14}S_5$, $Pd_{11}S_5$, PdS (palladous sulfide), and PdS_2 (palladium disulfide). These are dark and comparatively inert chemically, and they show some semimetallic characteristics.

Palladous nitrate, $Pd(NO_3)_2$, may be formed by dissolving finely divided palladium in warm nitric acid. The salt may be obtained as crystals from this solution, but it may be contaminated with basic salts and is very hygroscopic. The solution readily hydrolyzes, especially if heated.

RHODIUM

Rhodium shows a decided preference for the oxidation state +3 in its compounds. However, the oxidation state +4 is found in rhodium tetrafluoride, RhF_4; in a poorly characterized hydrous dioxide, $Rh(OH)_4$; and in a few fluoro- and chloro- complexes, such as cesium hexachlororhodate(IV), Cs_2RhCl_6. A sole instance of the oxidation state +6 is the hexafluoride, RhF_6. Oxidation states lower than +3 occur among the carbonyls and carbonyl halides and in a number of recently synthesized complexes. Rhodium is said to be the only element in the second or third transition series that possesses a definite, well-characterized aquo ion; this is the yellow rhodium hexaquo ion, $Rh(H_2O)_6^{3+}$ found in aqueous sulfate or perchlorate solutions.

The oxide, Rh_2O_3, results from heating the finely divided metal to red heat in air; it can also be prepared by igniting rhodium(III) nitrate, $Rh(NO_3)_3$. It is a gray crystalline solid with the same crystalline structure as corundum, and it does not dissolve in acids. When alkali is carefully added to rhodium(III) solutions, a yellow precipitate of hydrous rhodium(III) oxide, said to be $RH_2O_3 \cdot 5H_2O$, is formed. This will dissolve in acids or in excess alkali, and on ignition it forms the anhydrous oxide. If too much alkali is used, a black precipitate is produced that does not dissolve in acids; this is believed to be $Rh_2O_3 \cdot 3H_2O$.

Rhodium trifluoride, RhF_3, is produced by the action of fluorine on metallic rhodium or rhodium trichloride, $RhCl_3$, at 500–600° C. It is a red solid that is unreactive toward water, aqueous acids, and aqueous alkalis. This synthesis of rhodium trifluoride also results in the simultaneous production of a small amount of rhodium tetrafluoride, RhF_4, a blue solid.

Rhodium(III) chloride can be prepared in various ways, and its properties depend on the method of preparation. With direct union of the elements at 250° C, the product, $RhCl_3$, is a red powder insoluble in water and acids. If the yellow hydrous oxide is treated with hydrochloric acid and the solution carefully evaporated, dark-red crystals of rhodium chloride tetrahydrate, $RhCl_3 \cdot 4H_2O$, are produced; these are water-soluble. If these are dehydrated in a stream of gaseous hydrogen chloride, a water-soluble anhydrous salt is produced.

Several rhodium(III) salts containing the aquo ion, $Rh(H_2O)_6^{3+}$, have been prepared by dissolving the yellow hydrous oxide in the appropriate acid. The salts that have been characterized are formed by oxyacids, Rhodium perchlorate, $Rh(ClO_4)_3 \cdot 6H_2O$, is an example of such a compound, and its structure is known from x-ray diffraction. Rhodium alums, such as $KRh(SO_4)_2 \cdot 12H_2O$, have been known for a long time. Rhodium sulfate occurs in two forms: one is a yellow crystalline solid, $Rh_2(SO_2)_3 \cdot 14H_2O$; the other is red, $Rh_2(SO_4)_3 \cdot 6H_2O$. The yellow sulfate is a normal ionic salt from whose solutions barium sulfate can be precipitated. The red salt is obtained by evaporating the yellow solution to dryness at 100° C; from its solution, no barium sul-

Physical and Chemical Properties

fate can be precipitated. Evidently, in the latter compound, three sulfate ions are coordinated to the metal in a nonlabile complex.

IRIDIUM

Iridium exhibits a greater variety of oxidation states in its compounds than rhodium, its close congener in the periodic table. Among its simpler compounds, the preferred oxidation states are +3 and +4, with the former being more common. Some examples of compounds in which iridium displays other oxidation states are mentioned below, and a wide range of formal oxidation states have been encountered in its coordination complexes. In contrast with rhodium, iridium shows no evidence of an aqueous cation, so most of its solution chemistry involves complex ions.

Iridium dioxide, IrO_2, is the most common oxide and has the rutile structure. It is formed by direct union of the elements at about 1000° C, but it decomposes at about 1120° C. When quadripositive iridium salts (e.g., $IrCl_6^{2-}$) are treated with alkali, an intensely blue hydrous oxide is precipitated. When dried under nitrogen at 350° C, this yields the dioxide in a reasonably pure state. Addition of alkali to a terpositive iridium salt (e.g., $IrCl_6^{3-}$) in an oxygen-free atmosphere yields a green or blue-black hydrous sesquioxide. This is a gelatinous material, soluble in excess alkali and apt to absorb atmospheric oxygen with oxidation to the dioxide. If this preparation is dehydrated, even with oxygen excluded, it fails to yield a pure Ir_2O_3. The trioxide, IrO_3, has been prepared by fusion of the metal with alkaline oxidants, such as sodium peroxide; however, this compound remains poorly characterized.

Iridium hexafluoride, IrF_6, is a yellow solid (m.p., 44° C) that fumes strongly in air and reacts vigorously with water. The pentafluoride is also a very reactive yellow solid (m.p., 106° C). A third fluoride, IrF_3, is a black solid that is difficult to prepare.

Iridium trichloride, $IrCl_3$, is formed by direct union of the elements at 450–600° C, the reaction apparently being accelerated by sunlight. It is olive-green, brown, or black, according to particle size. It is not soluble in water. A hydrated iridium chloride, dark-green and water-soluble, is formed by reaction of hydrochloric acid on the dioxide. A tetrachloride of somewhat doubtful quality has been formed by the reaction of chlorine or aqua regia with $(NH_4)_2IrCl_6$ and by some other syntheses.

Iridium reacts with sulfur, selenium, and tellurium; the compounds formed have been identified as intermediate phases in the two-component systems with iridium. These include Ir_2S_3, IrS_2, Ir_3S_8,

IrS$_3$(?), Ir$_2$Se$_3$, IrSe$_2$, IrSe$_3$, IrTe$_2$, and IrTe$_3$. These are all dark solids and are quite resistant to acids.

RUTHENIUM

Ruthenium is known to occur in compounds in at least eight oxidation states, but the most common are +2, +3, and +4. In general, the chemistry of ruthenium resembles that of osmium much more than that of iron.

The tetroxide, RuO$_4$, has already been mentioned; it is a yellow molecular solid (m.p., 25° C; b.p., 100° C) and is highly toxic. It is produced when acidic solutions of ruthenium compounds are heated with strong oxidizing agents. In contrast with osmium tetroxide, it is not formed by direct union of the elements, nor is it produced by the action of nitric acid on ruthenium compounds. When ruthenium tetroxide is dissolved in alkali, it is immediately reduced first to a green perruthenate, RuO$_4^-$, and then to an orange ruthenate, RuO$_4^{2-}$.

The dioxide, RuO$_2$, is formed by heating ruthenium in air at 500–700° C. It is a black crystalline solid with the rutile structure. It is not dissolved by acids, but is reduced to the metal when heated in hydrogen. Addition of alkali to ruthenium(III) solutions results in precipitation of dark hydrous oxides; similar precipitates occur when alkaline solutions of ruthenium tetroxide are treated with ethanol and boiled. In neither case are the substances formed well characterized.

Ruthenium disulfide, RuS$_2$, is a gray-blue crystalline solid known in mineral form as laurite; it is structurally analogous to pyrite, FeS$_2$. It can be prepared by direct union of the elements at high temperature. The compound is chemically unreactive.

Treatment of the element with fluorine produces ruthenium pentafluoride, RuF$_5$, a dark-green solid (m.p., 85.6° C; b.p., 227° C) with a colorless vapor. It is very reactive, is hydrolyzed by water, and is reduced when heated with iodine. Treatment with an excess of iodine yields ruthenium trifluoride, RuF$_3$, a brown solid; but treatment with iodine and IF$_5$ results in formation of yellow crystals of the tetrafluoride, RuF$_4$.

Chlorination of the element yields ruthenium trichloride, RuCl$_3$, a black solid that is insoluble in water and of which there are two crystalline modifications. A hydrated form, prepared by evaporating a hydrochloric acid solution of ruthenium tetroxide in an atmosphere of hydrogen, is formulated as RuCl$_3 \cdot$H$_2$O. This is soluble in water, but

Physical and Chemical Properties

the fresh solution contains no chloride ion and should be regarded as a complex. This aqueous solution undergoes slow hydrolysis with precipitation of a hydrous oxide.

Ruthenium tetrachloride, $RuCl_4$, and a hydroxychloride, $RuOHCl_3$, are formed when hydrochloric acid solutions of ruthenium tetroxide are evaporated. There is reason to believe that these compounds are structurally more complex than the formulas suggest.

Perruthenates and ruthenates have been mentioned already in connection with the alkaline solutions of ruthenium tetroxide. Ruthenates and perruthenates may also be produced by direct fusion of the metal with a mixture of an alkali-metal nitrate and hydroxide. In the solid state, these substances are black; their aqueous solutions—green and orange, respectively—are not particularly stable.

OSMIUM

In compounds and complexes of osmium that have been described, it exhibits each of the nine oxidation states from 0 to +8. The more common values among its simpler compounds are +3, +4, +6, and +8. Like most of the platinum metals, osmium does not form a simple aqueous cationic species.

Osmium tetroxide, OsO_4, is a colorless molecular solid (m.p., 40° C; b.p., 101° C) with a characteristic pungent odor suggestive of ozone. It is a highly toxic substance, and exposure of the eyes and the respiratory tract to it must be avoided. It is formed directly by combination of the elements, e.g., when the metal is heated in air above 200° C. Also, and in contrast with ruthenium, the tetroxide is formed when any of many osmium compounds are heated with nitric acid. When osmium tetroxide is dissolved in water, it remains in the molecular form, but it is converted by alkali to the osmate ion, $HOsO_5^-$ or $OsO_4(OH)_2^{2-}$, which is yellow. It was undoubtedly this ion that early workers mistook for chromate before the identification of osmium.

A dioxide, OsO_2, can be prepared either by heating the metal in a stream of nitric oxide at 650° C or by heating osmium in a stream of nitrogen and osmium tetroxide vapor at 600° C. It is a dark solid, possibly dimorphic, with one form having the crystal structure of rutile. This oxide dissolves in hydrochloric acid to form chloroosmic acid(IV), H_2OsCl_6. Addition of alkali to the latter solution regenerates osmium dioxide in a hydrous form.

Osmium hexafluoride, OsF_6, is a yellow-green solid (m.p., 32.1° C;

b.p., 46° C) prepared by direct union of the elements at 250° C. In the chemical literature before 1958, this compound was erroneously described as OsF_8. It is readily reduced by iodine and can be hydrolyzed by water. A pentafluoride is formed from the hexafluoride by ultraviolet irradiation or by reduction by iodine dissolved in IF_5. It is a green solid (m.p., 70° C), melting to a blue liquid (b.p., 226° C) and yielding a colorless vapor. There is also a yellow tetrafluoride (m.p., 230° C) that can be prepared from the hexafluoride by reduction.

There is a good deal of contradictory information about the chlorides of osmium.[170] A red crystalline tetrachloride can be prepared by direct chlorination of the metal under a pressure of 7 atm (709 kN/m²). This dissociates when heated at 470° C in a stream of chlorine to form a dark-gray solid trichloride. The tetrachloride is soluble in water or alcohol, although the solutions are not particularly stable, whereas the trichloride is insoluble and not very reactive.

There are a disulfide, a diselenide, and a ditelluride; each can be formed by direct combination of the elements above 600° C. All are dark solids with the crystal structure of pyrite and show very little chemical reactivity.

Osmium and ruthenium form various oxy- compounds that have no counterparts among the other platinum metals, but these may be compared with the ferrates. The osmate(VIII) ion formed when osmium tetroxide dissolves in alkali has already been mentioned. Noteworthy is the observation that, when the tetroxide reacts with alkali, osmium, in contrast with ruthenium, suffers no immediate change in oxidation state. However, if a mild reducing agent (e.g., alcohol) is added to the alkaline solution, the element is converted to the osmate(VI) ion, $OsO_2(OH)_4^{2-}$, evident from the color (pink). Both osmate structures are octahedral about the metal atom. If aqueous osmate(VIII) is treated with concentrated aqueous ammonia, an unusual compound results, called an osmiamate, OsO_3N^-, in which a nitrogen atom is bound to the metal by what appears to be a multiple bond. Structurally, this ion appears to be a distorted tetrahedron; similar species are found for rhenium and molybdenum.

THERMODYNAMIC DATA ON BINARY ALLOYS

Considerable information regarding the platinum-group metals and their simpler compounds is available in compilations of thermodynamic data,[156,162,215] which should be consulted for further details. Information regarding binary intermetallic compounds formed as components in alloy systems can be found in such sources as the *Metals Handbook*.[189]

Physical and Chemical Properties

Coordination Compounds of Platinum and Palladium

The coordination chemistry of platinum and palladium has attracted considerable attention in recent years, largely because the metals have been the source of a large number of compounds of great intrinsic interest. Use of the metals and their compounds as homogeneous and heterogeneous catalysts has been the primary reason for the rapid development of the organometallic chemistry of these metals. Research in this subject is extensive and has resulted in a vast quantity of information on the reaction of the metals and their compounds with organic molecules. More recently, the discovery that *cis*-dichlorodiammineplatinum(II) exhibits anticancer activity (see Chapter 3) has stimulated tremendous interest in the effects of platinum compounds on biologic systems.[375]

Throughout the history and development of modern coordination chemistry, these elements have been of particular interest. The square-planar geometry of the bivalent oxidation states made possible the study of *cis* and *trans* isomers in such complexes. One of the most studied properties of these complexes is the labilizing effect that some ligands have at the *cis*[355] and *trans*[34] positions. Research has resulted in a better understanding of the mechanisms of substitution reactions involving metal complexes. The *trans* effect has also enabled the systematic synthesis of the geometric isomers of a given complex.[157]

OXIDATION STATES AND STEREOCHEMISTRY

For the different oxidation states of platinum and palladium, the partially filled shells are d shells—$5d$ and $4d$, respectively. These d orbitals project well out to the surface of the atoms and ions, so the electrons occupying them are strongly influenced by the surroundings of the ion and, in turn, are able to influence their environment significantly. Thus, many of the properties of a particular oxidation state are quite sensitive to the number and arrangement of the d electrons present. For this reason, the coordination number and stereochemistry for the individual oxidation states are different.

The most common oxidation state of both metals is $+2$.[186] Almost all the complexes of this oxidation state have a coordination number of 4 and a square-planar geometry. There are also many compounds with the elements in the 0 and $+4$ oxidation states. However, these oxidation states are much more common for platinum than for palladium. In the $+4$ oxidation state, both metal ions are in an octahedral ligand environment. Zerovalent platinum and palladium complexes have coordination numbers between 4 and 2, with 4 being the most common. The 4-coordinated, zerovalent complexes are tetrahedral, inasmuch as in these complexes the electronic configuration of the metals is d^{10}, rather than d^8s^2. Compounds in which platinum and palladium have oxidation states other than 0, +2, and +4 are rare. There are a few compounds of platinum with oxidation states of +1, +5, +6, and possibly +3; the only other oxidation state of palladium is apparently +1.

COMPLEXES OF THE ZEROVALENT METALS

Zerovalent platinum and palladium form complexes with phosphine,[265] arsine,[265] phosphite,[265] isocyanide,[266] cyanide,[70] acetylide,[290] ammonia,[465] and nitric oxide[171] ligands. Carbon monoxide forms only a polymeric complex with platinum(0) in the absence of other ligands. However, it forms a variety of monomeric and polymeric complexes with phosphine complexes of platinum(0). The strong σ-donor ability of phosphine increases the electron density on the metal, making it more susceptible to metal-to-carbon monoxide π-back-donation.[266,465]

Most of the preparative routes involve reduction of complexes of the bivalent metals in the presence of the ligand to be complexed.[70,265] However, in a few instances, complexes have been prepared by heating the metal in the presence of the ligand. For example, [Pd(PPh$_3$)$_2$] can be formed by heating metallic palladium with triphenylphosphine in the presence of excess triethylsilane.[181]

The zerovalent complexes of platinum and palladium that have been most widely investigated are those containing tertiary phosphine ligands. These complexes are air-stable in most instances and are soluble in a number of common organic solvents. For this reason, they are used as the starting material for the synthesis of many of the other zerovalent complexes. Some of the properties of these complexes are shown in Table 4–6.

COMPLEXES OF THE BIVALENT METALS

Bivalent platinum and palladium form complexes with ligands containing donor atoms from almost every group in the periodic table. In the following discussion, some of the more important complexes of several groups are briefly described.

- *Hydride Complexes of the Bivalent Metals:* The first hydride complexes of platinum and palladium were reported in 1957.[91] There has since been a rapid expansion of the field, resulting in a number of

Physical and Chemical Properties

TABLE 4-6 Properties of Zerovalent Platinum and Palladium Complexes

Complex	Color	Melting Point, °C	Remarks
[Pt(PPh$_3$)$_4$]	Yellow	118 (decomp.)	Stable in air for several hours
[Pt(PPh$_3$)$_3$]	Yellow	125–130 (decomp.)	Stable in air for several hours
[(PPh$_3$)$_2$PtO$_2$]	Yellow	130–132	Stable in air
[(PPh$_3$)$_2$Pt(C$_2$H$_4$)]	White	122–125	Stable in air
[Pd(PPh$_3$)$_4$]	Yellow	100–105 (decomp.)	Stable in air for only short period

reviews on the topic.[88,114] A wide range of methods have been devised for their preparation. A typical synthesis[95] is the reaction of *cis*-[(PR$_3$)$_2$PtCl$_2$] with hydrazine in dilute aqueous or alcoholic solution to produce *trans*-[(PR$_3$)$_2$PtHCl]. All the known hydride complexes are square-planar, and most have a *trans* configuration. The platinum complexes are considerably more stable than the palladium compounds. For example, *trans*-[(PEt$_3$)$_2$PtHCl] can be distilled under high vacuum at 130° C and 0.01 torr (1.33 N/m²),[95] whereas the palladium compound is isolated as a solid and is always contaminated with decomposition products.[92] The properties of both the metal and the ligand that produce a high ligand-field stabilization energy in the complex also produce a hydride complex that is more stable with respect to air and water. One of the most interesting and important reactions of the hydrides involves insertion of an unsaturated organic compound into the metal–hydride bond.[90] Reactions of this type are involved in the homogeneous hydrogenation and hydrosilation of olefins, as well as the catalytic isomerization of olefins by both platinum(II) and palladium(II) complexes.

• *Complexes with Group IVA Elements:* There are many complexes in which ligands containing silicon, germanium, tin, and lead bond directly to either platinum(II) or palladium(II). Carbon is also in this group, but complexes involving the metal-to-carbon bonds usually have different properties. There are several reviews on complexes containing bonds between metals and group IVA elements.[61,93] A number of methods have been used to prepare these complexes.[44,93] Treatment of *cis*-[(PEt$_3$)$_3$PtCl$_2$] with (R$_3$X)$_2$Hg (X = silicon, germanium, and palladium; R = organic group) has been used to prepare *trans*-[(PEt$_3$)$_2$

PtCl(XR$_3$)] in benzene. The complexes of platinum and palladium that have been prepared in the pure state are all crystalline. It appears that platinum complexes are more stable than the corresponding palladium complexes; the order of stability of the group IVA ligands is Cl$_3$X > Ph$_3$X > Me$_3$X. The order of stability of the platinum complexes is approximately Sn ~ GE > Si ~ Pb; for the palladium complexes, it is Ge > Pb >> Sn ~ Si. The presence of tertiary phosphine or arsine ligands appears to be virtually essential for the formation of stable complexes in which the group IVA elements, including carbon, are bound to platinum or palladium.

- *Complexes with Group VA Elements:* Some of the most interesting complexes of platinum(II) and palladium(II) are those involving ligands that contain the group VA donor atoms nitrogen, phosphorus, arsenic, and antimony. Nitrogen in almost any environment binds strongly to both metals, and this has led to the preparation of complexes with a wide range of nitrogen ligands. Tertiary phosphine ligands have been very important in the development of the organometallic chemistry of platinum and palladium through their ability to form very stable complexes with both metals. All four group VA elements form strong σ-bonds to the metals. Phosphorus, arsenic, and antimony are also capable of accepting back-donations of electron density from the metal through π-bonds. This accounts for the considerable difference in properties between the nitrogen-containing ligands and ligands containing the heavier group VA element.

Probably the most important nitrogen-containing complexes of platinum(II) and palladium(II) are the bisammine complexes, [PtL$_2$X$_2$] (L = ammonia or amine; X = halogen). These complexes exist as *cis* and *trans* isomers and have been very instrumental in the development of modern coordination chemistry. Studies on these complexes also led to the discovery of the *trans* effect in platinum(II) complexes in the 1920s.[96] More pertinent to this discussion are the biologic effects of *cis*-dichlorodiammineplatinum(II) recently observed. This complex has been found to be a potent anticancer agent and is rather toxic.

The tertiary phosphine, arsine, and stibine complexes of platinum(II) and palladium(II) are stable in air, are soluble in organic solvents, are readily recrystallized, and have well-defined melting points. The stability of the compiexes decreases in the order PR$_3$ > AsR$_3$ > SbR$_3$.[57] The monomeric platinum(II) complexes generally exist as both *cis* and *trans* isomers. The phosphine and arsine ligands are always *trans*.[57] The corresponding stibine complexes contain the *cis* isomers at up to 40% in equilibrium in solution.

Physical and Chemical Properties

• *Bivalent Complexes with Group VIA Elements:* Platinum(II) and palladium(II) form stronger complexes with ligands containing sulfur than they do with oxygen-containing ligands. The stabilities of complexes of sulfur, selenium, and tellurium are very similar. The actual sequence depends on the nature of the other ligands bound to the metal. The relative stability of complexes involving oxygen-containing ligands is $H_2O > ROH > R_2O$; the opposite is true for sulfur-containing ligands: $H_2S < RSH < R_2S$. Complexes of bivalent platinum and palladium containing H_2X (X = group VIA element) are known only when X is oxygen. The compound $[Pd(H_2O)_4]^{2+}$ $(ClO_4)_2$ has been prepared, but the aquo complexes are rarely isolated, because water is a very poor ligand for these metal ions. Hydroxy complexes are known, but are also rarely isolated. The corresponding HX^- complexes involving the other elements of group IVA cannot be prepared. However, thioethers, selenoethers, and telluroethers all form complexes with both metals. The telluroether complexes of platinum(II) are much less stable than their sulfur and selenium analogues. Other types of ligands that contain group VIA elements and form complexes with platinum(II) and palladium(II) are sulfate ion,[128] carbonate ion,[305] nitrate ion,[257] urea,[318] thiourea,[46] and many organic polydentate ligands.[186]

COMPLEXES OF THE QUADRIVALENT METALS

Platinum readily forms platinum(IV) compounds and complexes, whereas palladium is reluctant to form palladium(IV) compounds. The palladium complexes are more stable than simple palladium(IV) compounds, but only a few are known; apart from the complexes formed on dissolution of palladium in concentrated nitric acid, they are mainly the octahedral halide anion complexes.

None of the group IVA elements form complexes with palladium(IV). Platinum(IV) forms cyanide, cyclopentadienyl, and many organometallic complexes. Stable platinum(IV) complexes with other group IVA ligands have been isolated by the addition of hydrogen halides to platinum(II) complexes that have ligands containing these elements.

Platinum(IV) forms a large number and a wide variety of complexes with nitrogen-, phosphorus-, and arsenic-containing ligands. There are very few palladium(IV) complexes with ligands involving these donor atoms. The most extensive and typical series of platinum(IV) complexes are those which span the entire range from the hexammines, $[Pt(amine)_6]X_4$, to the hexahalide complex, $[PtX_6]^{2-}$.[444] Other complexes of the quadrivalent metals with group VA elements are tertiary phosphine,[89] tertiary arsine,[303] alkylcyanide and arylcyanide,[164] and azide[40] ligands.

There are very few complexes of palladium(IV) that contain ligands with group VIA donor atoms. Platinum(IV), however, forms a much wider range of complexes, with such ligands as dialkylsulfides, dialkylselenides, nitrate, carbonate, and sulfite.

ORGANOMETALLIC COMPLEXES

Several reviews have been written on the organometallic chemistry of platinum and palladium.[304,315] Both platinum and palladium form complexes containing metal–carbon σ-bonds. Complexes of this type with both platinum(II) and platinum(IV) are known, whereas palladium forms such complexes only in the +2 oxidation state. Complexes of these metals involving metal–carbon σ-bonds contain other ligands, such as phosphines,[76] arsines,[76] bidentate thioethers,[76] selenoethers,[401] bipyridyl,[76] pyridine,[243] and triethylstibine.[474] Treatment of a halometal complex with an anionic alkylating agent, such as a Grignard or organolithium reagent, is the most widely used method for preparing both the platinum(II) and palladium(II) complexes.[76]

The platinum(II) alkyl and aryl complexes are colorless crystalline solids that are not oxidized by moist air. The palladium(II) complexes are less stable. Both metals form a wider variety of organometallic complexes than any other transition metal. When the complexes are heated, they appear to undergo homolytic fission, with the formation of organic radicals. For example, when $[(PEt_3)_2Pd(CH_3)_2]$ is heated at 100° C in a sealed tube, a mixture of ethane and ethylene is obtained, with a trace of methane.[76]

Olefin complexes of platinum are the oldest class of organometallic complexes known and have therefore been extensively investigated. A number of their reactions are of commercial interest.[412] Platinum and palladium form olefin complexes in both the bivalent and zerovalent oxidation states. One method of preparing the platinum(II) and palladium(II) complexes is to react the metal(II) salts, such as $K_2[PtCl_4]$, with the olefin in aqueous or nonaqueous solution.[10] Platinum(II) olefin complexes are much more stable than palladium(II) olefin complexes, with respect both to displacement of the olefin with a halogen and to reactions of the complex with a variety of reagents, including water and most nucleophiles. The olefin and acetylene complexes of the zerovalent metals are stable, generally white, crystalline materials with fairly high decomposition points. The decomposition temperature for one of the several complexes is given in Table 4–6. The importance of

Physical and Chemical Properties

electron-withdrawing substituents attached to the multiple bond is indicated by the decreasing thermal stability of $[(PPH_3)_2 Pt(R_2C=CR_2)]$ in the order R = CN > F > H.

Another important class of organometallic complexes of platinum and palladium is the π-allyl complexes. The difference between these complexes and normal olefin complexes is that a three-carbon π-allyl group is involved in the bonding of the former. Palladium(II) forms a wide range of π-allyl complexes, whereas platinum forms very few. Several methods have been reported for the synthesis of these complexes.[259,390,402,479] The platinum and palladium π-allyl complexes are yellow or occasionally red crystalline complexes, most of which are easily handled in air at room temperature. The palladium complexes are hydrolyzed readily at room temperature in the presence of water.

Complexes of Ruthenium

The chemistry of ruthenium has been extensively reviewed by Griffith.[170] A wide range of complexes are known for ruthenium in the 0, +2, +3, and (less commonly) +4 oxidation states. The largest number and variety of complexes are found for ruthenium(II). Extensively studied subjects involve the chemistry of trialkylphosphines and triarylphosphines, the corresponding phosphites, and, to a lesser extent, the arsines. Other ligands often associated with the PR_3 group in ruthenium complexes are halogens, hydrogen, alkyl and aryl groups, carbon monoxide, nitric oxide, and alkenes. Some typical complexes containing PR_3 ligands are $RuCl_2(PPH_3)_3$, $RuCl_2(CO)(PPh_3)_2$, $RuH_2(N_2)(PPh_3)_3$, $RuCl(NO)(PPh_3)_2$, and $RuCl_2(RCN)_2(PPh_3)_2$.

The chemistry of nitrogen-donor ligands is of special interest, owing to formation of N_2 complexes of ruthenium(II). The first nitrogen complex to be prepared was $[Ru(NH_3)_5N_2]Cl_2$,[11] which is a frequent precursor for the synthesis of many other ruthenium(II) ammine complexes. The hexammine can be prepared by reduction of zinc dust with strong ammoniacal solutions of ruthenium halides.[254] The $Ru(NH_3)_5^{2+}$ group has strong π-bonding properties and forms complexes with nitrogen, carbon monoxide, and similar π-bonding ligands. The aquopentammine reacts with both nitrogen[25] and nitrous oxide[26] by displacing the water molecule. Ethylenediamine and other organic ammines also form a wide variety of complexes with ruthenium(II).

Ruthenium(III) forms several types of ammine complexes. The reduction of ruthenium ammines of the type $[Ru(NH_3)_5L]^{3+}$ with Cr^{2+} and other reducing agents has been studied in detail.[285] The hexammine, $[Ru(NH_3)_6]^{3+}$, reacts with water very slowly at room temperature, but reacts rapidly with nitric oxide to form $[Ru(NH_3)_5NO]^{3+}$. Another type of ammine complex formed by ruthenium(III) is "ruthenium red." The structure of "ruthenium red" appears to be that of a linear trinuclear ion with oxygen bridges between the metal atoms, $[(NH_3)_5Ru-O-Ru(NH_3)_4-O-Ru(NH_3)_5]$.

The formation of nitric oxide complexes is a characteristic feature of ruthenium chemistry. The vast majority of ruthenium–nitric oxide complexes are of the general type $[Ru(II)(NO)L_5]$ (L = almost any ligand).

In addition to the complexes already discussed, there are several well-characterized oxygen-ligand complexes of ruthenium, such as oxalates, $[Ru(ox)_3]^{3-}$, and acetylacetonates, $Ru(acac)_3$. Many sulfur-donor complexes are also known.[94]

Complexes of Rhodium and Iridium

The chemistry of rhodium and iridium has been thoroughly treated by Griffith.[170] Most of the chemistry of rhodium involves the oxidation states -1, 0, $+1$, and $+3$. The most common oxidation states of iridium are $+1$, $+3$, and $+4$.

The coordination chemistry of rhodium and iridium in the $+1$ state primarily involves π-acid ligands, such as carbon monoxide, PR_3, and alkenes. Both square-planar and 5-coordinate species are known for both elements. The rhodium(I) and iridium(I) complexes are usually prepared by some form of reduction of $RhCl_3 \cdot 3H_2O$ or K_2IrCl_6 in the presence of the complexing ligand. Some typical rhodium(I) complexes are $[Rh(CO)_2Cl]_2$, trans-$RhCl(CO)(PPh_3)_2$, $RhH(CO)(PPh_3)_3$, and $RhCl(CO)(PPh_3)_2$. The complex $RhH(CO)(PPh_3)_3$ undergoes a wide range of reactions, but its main importance is as a hydroformylation catalyst for alkenes.[62] Chlorotris(triphenylphosphine)rhodium, $RhCl(PPh_3)_3$, is widely used as a homogeneous hydrogenation catalyst. $RhCl(PPh_3)_3$ also undergoes a wide range of oxidative-addition and other reactions.[31,50,445] The most important iridium(I) complexes are trans-$IrCl(CO)(PPh_3)_2$ and its analogues containing other phosphines. Many studies involving their oxidative-addition reactions have been reported.[457]

Both rhodium(III) and iridium(III) form a large number of cationic, neutral, and anionic octahedral complexes. The cationic and neutral complexes of both elements are generally kinetically inert, but the anionic complexes of rhodium(III) are usually labile. Both rhodium and iridium form cobalt-like ammines of the types $[ML_6]^{3+}$, $[ML_5X]^{2+}$, and $[ML_4X_2]^+$. The salts are made in various ways, but usually by the

Physical and Chemical Properties

interactions of aqueous solutions of $RhCl_3$(aq) with the ligand. Some typical cationic complexes are $[Rh(NH_3)_5Cl]^{2+}$, trans-$[Rhpy_4Cl_2]^+$, trans-$[Irpy_4Cl_2]^+$, and $[RhH(NH_3)_4H_2O]^{2+}$. Neutral complexes with carbon monoxide, PR_3, pyridines, etc., as ligands are prepared directly from $RhCl_3 \cdot 3H_2O$ or $IrCl_6^{3-}$. Typical examples of some neutral complexes are $RhCl_3L_3$, $IrCl_3py_3$, and $Rh(acac)_3$.

Complexes of rhodium(II) are rare and are confined primarily to binuclear carboxylates of the type $[Rh(OOCR)_2]_2$. The complexes have a tetra-bridged structure that involves a Rh–Rh bond. These compounds are effective antitumor agents when given to tumor-bearing mice[137] and are potent enzyme-inhibitors.

Multimetallic Cluster Compounds

Strictly speaking, "cluster compounds" are defined [186] as discrete units containing three or more transition elements of the same or different types in which strong metal–metal bonds are present. However, the term is also used often to include complexes in which at least two transition-metal atoms are held together by bridging species, including carbonyls, triphenylphosphines, and sulfur species. Several species include platinum-group metals. Most of these complexes are covalent structures and are relatively insoluble (and may be unstable) in water.

Considerable interest has been manifested in these complexes recently, because they offer the possibility of varying the intermetallic distances in zerovalent systems. This feature makes them promising candidates for highly selective catalytic reactions.

The physiologic activity of these bridged cluster compounds is not known. Although they may not be very soluble in water, they may be soluble in lipids, and this could allow them to be transported easily into living cells. As interest in these species expands, it is recommended that they be tested for toxicity.

5

Analysis and Determination

The analytic chemistry of the platinum-group metals (and gold) has been surveyed in two volumes compiled by Beamish and Van Loon,[35,37] which provide an authoritative source for much of the material in this chapter. There is also an excellent review by Walsh and Hausman[162] that surveys the state of the art up to about 1962. Reports from a recent symposium on the analytic chemistry of the platinum-group metals have been published.[19]

This chapter covers methods of sampling the platinum-group metals and bringing them into solution, appropriate methods of separating them from each other and from base metals, and various "wet-chemical" and instrumental methods for their determination.

SAMPLE PREPARATION

The sampling of various materials containing platinum-group metals is difficult, because such items are often not homogeneous. This is particularly true of catalysts, which may be in the form of powder, pellets, or rods containing a wide variety of contaminants after use. Many catalysts are concentrated on the surface of supports, so fines are richer than the bulk in metal concentration. Care must then be taken to sample both fractions and to blend each individually; in the case of catalyst powders, this requires utmost care in thorough homogeniza-

Analysis and Determination

tion. Catalysts containing organic matter and those on carbon supports can be burned and the residue prepared for sampling by grinding, sieving, and blending. However, because of the possible risks in burning residues containing flammable and catalytic material, a small portion should be treated first as a guide in selecting the proper and safe procedure.

Decomposition and dissolution of the platinum metals present special problems, because of the general inertness of these "noble" metals.

Behavior of Mineral Acids

Although the platinum metals are referred to as "noble"—i.e., not readily attacked by single mineral acids or by oxygen—some are corroded to an extent that depends on the presence of impurities or on the degree of subdivision or dispersion in another metal (see Chapter 4). For example, the use of nitric acid to remove copper from a platinum precipitate can result in losses of the latter into the acid copper solution. Similarly, the presence of mercury will permit nitric acid to dissolve some platinum. In fire-assay methods (described below), in which the platinum metals are isolated in a "button" of lead alloy, platinum and rhodium are attacked appreciably if the button is dissolved in nitric acid.

Aqua regia, a mixture of three or four parts of hydrochloric acid and one part of nitric acid, is well known as a dissolving agent for gold and platinum. However, not all the platinum metals are quantitatively dissolved by this reagent (see Chapter 4). The solvent role of aqua regia depends on the presence of nitric acid—a powerful oxidizing agent capable of electron removal, e.g., $Pt^o = Pt^{4+} + 4e$—and hydrochloric acid, which provides a source of chloride ions to form a stable complex with the oxidized metal ions, namely, $PtCl_6^{2-}$. Other combinations of oxidants and complexing agents will also have great dissolving power; for instance, air and potassium cyanide solution will dissolve even gold.

Finely divided precipitates of platinum or palladium may be dissolved in aqua regia; but, if such solutions are evaporated, care must be taken not to overheat the residues, inasmuch as palladium may consequently become extremely resistant to redissolving, even by aqua regia. Massive platinum, as in heavy-gauge wire, is sometimes incompletely dissolved by aqua regia, and other methods must be used. Finely divided rhodium is soluble in hot, concentrated sulfuric acid; but, because of the stability of the rhodium sulfate complex formed, the metal is precipitated from this solution as sulfide, which is then dissolved in hydrochloric acid. No combination of mineral acids can dissolve ruthenium, iridium, or osmium.

Behavior of Alkalis

Fusion of the more intractable platinum metals with potassium hydroxide and either potassium nitrate or peroxide is usually effective in bringing them into soluble forms. By such means, rhodium, ruthenium, iridium, and osmium can be completely oxidized and later dissolved in mineral acids. Such fusion mixtures react with palladium, but less readily than the metals listed above. Caustic fusion is not applicable to the dissolving of platinum for analytic purposes, but may corrode platinumware.

Action of Chlorine

All the platinum metals can be solubilized by direct chlorination at high temperatures (650–700° C) in the presence of an alkali chloride, or by heating in a sealed reaction tube containing hydrochloric acid and an appropriate oxidant.

Fusion with Base Metals

Fusion of the more resistant platinum metals with zinc, tin, or bismuth produces alloys or intermetallic compounds that are more readily soluble in mineral acids or oxidizing fusions. Such treatment is particularly effective for larger pieces of the metals. Indeed, fusion of platinum with zinc may result in finely divided metal that reacts violently on addition of acids.

Fusion with lead can be used as a practical means of separation of the platinum metals into two groups: platinum, palladium, rhodium, gold, and silver dissolve in molten lead; iridium, ruthenium, and osmium form a separate, extremely insoluble crystalline phase. However, some analytic chemists have found that part of the iridium, ruthenium, and osmium remains associated with the lead phase; thus, this separation technique is questionable.

Pyrosulfate Fusion

Fusion with potassium pyrosulfate at 700° C in a quartz crucible is particularly useful for materials containing rhodium, such as

Analysis and Determination

palladium-rhodium alloys, which are not ordinarily soluble in aqua regia.

SEPARATION OF THE PLATINUM-GROUP METALS

In recent years, a number of new and more efficient methods for separating the platinum metals have been developed—such as ion exchange, solvent extraction, and chromatography—in addition to the more traditional techniques of fire assay and precipitation. In many respects, fire-assay methods, which involve a general technique for removing the platinum metals from ores or alloys to a medium favorable for analysis, are superior to some of the more modern techniques. Separation by selective precipitation (gravimetric determination) is discussed later. It should be noted that ruthenium and osmium, which form low-boiling tetroxides (see Chapter 4), can be readily separated from the other metals by volatilization.

Fire Assay

Methods included under the classification of "fire assay" involve high-temperature treatment of samples intimately mixed with graphite, an oxide of lead or copper, and a flux of borax and sodium carbonate, to produce an alloy containing the noble metals (including gold and silver) and a glassy slag containing all the extraneous materials. The slag is easily separated from the alloy button on cooling. If lead is used, the button can be fired again in air on a bone-ash cupel. Lead oxide (litharge) is formed and, being molten at the temperatures used, soaks into the porous cupel, carrying most of any remaining base metals with it. The precious metals remain as a small bead. Some loss of platinum metals—such as ruthenium, osmium, and iridium—may occur in cupellation, so direct dissolving of the lead or copper button is usually preferable.

Solvent Extraction

Removal of one member of the group with an organic solvent provides a useful, convenient, and selective separation technique. Of many schemes that have been proposed, two general categories may be distinguished: chelate extraction systems in which the platinum metal is incorporated into a neutral organic complex that is preferentially dissolved in an organic phase, and ion-association systems in which the platinum metal is in an anionic complex species (e.g., halide or pseudohalide, such as thiocyanate and complexes) that couples with a suitable cation (such as quaternary ammonium, phosphonium, or arsonium ion or a cationic dye) to form a neutral extractable species. Many of these complexes are characteristically and intensely colored, so the metal may be spectrophotometrically determined at low concentrations in the extract.

For example, platinum and palladium may be selectively separated from most of the other metals of the group by extraction with dipyrrolidinodithiocarbamate in chloroform. Iridium is separated from platinum, palladium, and rhodium by extraction with diantipyrylpropylmethane. A mixture of ruthenium, rhodium, and palladium in hydrochloric acid solution can be treated with a chloroform solution of triphenylmethylarsonium chloride to remove palladium; further treatment of the aqueous solution with 8-quinolinol in butyl Cellosolve removes ruthenium, leaving rhodium in the aqueous layer.

Chromatography

Chromatographic methods involve multistage countercurrent contact between an immobilized phase (liquid or solid) and a mobile phase (liquid or gas). This general category encompasses processes involving ion-exchange mechanisms, as well as solvent extraction (called partition chromatography). The immobilized phase is dispersed in a relatively thin layer supported by (or forming part of) finely divided granules that can be packed in a columnar bed or spread thinly on a glass or metal plate or even on papers of various sorts.

Selective absorption or binding of ions of interest from the sample solution when it is passed through a column packed with a suitable ion-exchanger constitutes a very convenient separation scheme. As mentioned previously, the platinum metals form many anionic halide or pseudohalide complexes; anion-exchange resins are used in their separation. The anionic complexes can be eluted from the column with either concentrated hydrochloric acid or 2 M perchloric acid. Cation-exchange resins have been used to remove base metals, as cations, from solutions that contain platinum-metal complexes as anions.

Paper chromatographic procedures have been developed for separating and qualitatively testing for the platinum metals with systems based on the extraction of halide anionic complexes of the metals into butanol.

Analysis and Determination

METHODS OF DETERMINATION

Spectrochemical Methods

EMISSION SPECTROSCOPY

Emission spectroscopy consists of excitation of samples with an arc or a spark under carefully controlled conditions so as to generate characteristic spectral lines whose intensity, recorded on photographic plates or measured by a photomultiplier tube, is related to the concentration of elements in the samples. This is a well-established, sensitive, and convenient way to analyze minor or trace constituents.

Such methods have been developed for the platinum-group metals, and some thoroughly proven procedures have been selected for publication by the American Society for Testing and Materials.[17] For this application, emission spectroscopy has been generally preferred by European analysts to atomic-absorption techniques, whereas the reverse is true in North America and Japan. Prior separation or concentration steps to bring the element sought to a suitable degree of detectability may be required with both techniques. And there is a requirement for the preparation of standards for calibration; not only must these contain the sought element(s) at known concentration (and usually a reference element at fixed concentration to serve as an internal standard), but, because of so-called matrix effects, they often must contain about the same proportions of other elements as do the samples for analysis.

The limits of detectability of the metals in a direct-current arc are: palladium, 3 ppm; platinum, 10 ppm; rhodium, 3 ppm; iridium, 100 ppm; ruthenium, 80 ppm; and osmium, 500 ppm. Similar limits are obtained with a high-voltage spark.

ATOMIC-ABSORPTION SPECTROSCOPY

In atomic-absorption spectroscopy (AAS), the metals of interest are converted to atoms in the vapor phase. Light of a wavelength characteristic of the sought element is then passed through the gas, and the amount that is absorbed by these gaseous atoms is measured and related to the amount present. Formation of the gaseous element can be achieved either by aspiration of a solution containing metal ions into the reducing part of a flame or (more recently) by flameless reduction of the sample heated in a carbon furnace.[138] Another fairly well-developed method of improving sensitivity, as well as the selectivity, when flames are used involves aspiration of extracts of the sample into organic solvents, rather than aqueous solutions into the burner.

A somewhat related technique, atomic-fluorescence spectroscopy (AFS), differs from AAS in that the atoms of the element(s) to be determined are subjected to excitation through illumination, but then lose part of the excitation energy through fluorescence emission. The latter is measured by a sensitive detector as with AAS, except that the beam of incident radiation is arranged at right angles to the direction along which the fluorescence is measured.

These spectroscopic techniques are susceptible to interferences resulting from substances in the sample that may interfere with the observed final measurement. Such interferences may be physical, such as alteration of the aspiration or nebulization of the analytic solution; or they may be chemical, such as something that influences the population of free atoms of the analyte(s). These effects call for careful control of the preparation of samples for vaporization; they may be alleviated to some degree by addition of releasing or protective agents (sometimes called "buffers") to the analyte. Such interferences are quite pronounced in the analysis of some of the platinum metals. For example, the addition of lanthanum can increase apparent spectral sensitivity in the determination of rhodium.

Examples of applications to the platinum metals, taken from a recent monograph,[242] illustrate the sensitivity of the method. It may be added that somewhat lower values obtained for these elements by the use of flameless excitation have been reported recently.[7] Most recent publications indicate a preference for this method of atomization.[275,441] See Table 5-1 for sensitivities.

TABLE 5-1 Atomic Spectroscopy of Platinum-Group Metals[a]

Element	Atomic-Absorption Spectroscopy		Atomic-Fluorescence Spectroscopy
	Sensitivity, ppm absorption/%	Detection Limit, ppm	Detection Limit, ppm
Iridium	7.7	—	4
Osmium	1–3	—	—
Palladium	0.2	0.02	0.06
Platinum	2.2	0.1	0.15
Rhodium	0.35	0.03	10
Ruthenium	0.25	—	80

[a] Data from Kirkbright and Sargent.[242]

Analysis and Determination

INDUCTIVELY COUPLED PLASMA SPECTROSCOPY

A development of considerable importance in spectroscopic analysis has been the introduction of high-frequency plasmas[58,139,140,250] as a means of atomic excitation. These attain much higher temperatures than flames and furnaces. The special advantage claimed for this technique is the capability of performing simultaneous multielement determinations in the parts-per-billion range with small volumes of sample. This represents a significant extension of the range and versatility of conventional emission spectroscopy.

Recently published detection limits for palladium, platinum, and rhodium were 0.007, 0.08, and 0.003 ppm, respectively, with this technique.

SPECTROPHOTOMETRY[391]

The platinum metals form highly colored complexes with many inorganic and organic reagents, and the literature[37] contains hundreds of spectrophotometric procedures. Only a few of the most widely used are mentioned here.

Platinum (IV) in hydrochloric acid reacts with tin(II) chloride to form rapidly a stable yellow-orange complex (403 nm)* that is useful for accurate determination of platinum at 3–25 ppm. The sensitivity can be enhanced by extraction of the complex into oxygen-donor solvents. A further increase in sensitivity might be achieved by using the near-ultraviolet absorption peak at 310 nm.[462] Gold and palladium interfere strongly, and the other platinum metals must be absent for highest accuracy.

5-(*p*-Dimethylaminobenzylidene)-rhodanine, which can also serve as a reagent for palladium, can be used for platinum at 0.5–6 ppm. A dark-red complex (545 nm) is developed in acid solution. Silver, gold, copper, rhodium, and palladium interfere. The addition of dimethylglyoxime permits a 12-fold excess of palladium to be tolerated.

2,3-Quinoxalinedithiol forms colored complexes with both platinum (624 nm) and palladium (517 nm), with the possibility of simultaneous determination of both metals.

Palladium forms a complex (594 nm) with pyridine-2-aldehyde-2-quinolylhydrazone that can be extracted into chloroform and used to determine the metal at 1.5–7 ppm in the presence of relatively large proportions of associated noble and base metals. Palladium can also be determined at 0.1–10 ppm by means of the intensely colored red-brown complex PdI_4^{2-} (410 nm) formed with excess iodide.

*Wavelengths for maximal absorption shown in parentheses.

Rhodium forms a cherry-red complex (510 nm) with p-nitrosodimethylaniline, useful at 0.15–1.1 ppm; however, associated base metals and platinum metals interfere. p-Nitrosodiphenylamine can also be used; its decreased sensitivity is compensated for by requiring less rigid control of conditions used for color development. Rhodium can be determined in the presence of platinum, copper, and iridium by the colored complex formed with tin(II) chloride after separation of platinum as an iodide complex into tributylphosphate in hexane.

Ruthenium forms a red complex (510 nm) with 2,4,4-tri-(2-pyridyl)-2-triazine at a pH of 2–4; this complex can be used for the determination of ruthenium at 1–4 ppm. Distillation of ruthenium beforehand is usually necessary, because of interferences.

Iridium may be determined in the presence of platinum, palladium, and rhodium by extraction with thenoyltrifluoroacetone in the presence of ethylenediaminetetraacetate (EDTA), with measurement of absorbance of the extract at 440 nm. The preferred range of concentration of metal in the extract for measurement in a 1-cm cell is 7–35 ppm.

Osmium forms a blue complex (600 nm) with thiocyanate, which, when extracted into ether or octanone, can be used for determinations of this metal at up to 15 ppm without interference, except from platinum and antimony.

X-RAY FLUORESCENCE SPECTROSCOPY

X-ray excitation of the elements of higher atomic number (>20) results in the production of characteristic fluorescence that provides the basis for highly selective, sensitive, rapid, and nondestructive analytic methods. This method has been applied to the determination on active alumina catalysts of platinum and palladium at 0.25–0.75%, with a standard deviation (at the 0.6% level) reported as 0.006% and 0.0025% by different workers.[286] In another instance, the same metals at approximately 3 and 4 ppm in a nickel matte were determined, with results comparing favorably with those of other methods. For this second application, the precious metals were concentrated by absorption on filter paper impregnated with ion-exchange resin, and determinations were made by comparison with standards. The other platinum metals have also been determined by x-ray fluorescence.

SPARK-SOURCE MASS SPECTROMETRY

Mass-spectrometric techniques have been extended by the use of an electric spark[8] to substances not readily volatilized. This multielement

Analysis and Determination 75

technique has been used to determine nickel, copper, lead, silver, and palladium as impurities in platinum at 0.5–14 ppm with reasonable accuracy. The procedure requires preliminary separation by ion exchange, with electrodeposition of these metals onto a gold electrode, from which they can be volatilized. This particular application is based on isotopic dilution.

SURFACE ANALYSIS

Three new techniques, each applicable to the analysis of elements at sample surfaces, are in various stages of development: ion-scattering spectrometry, scattered-ion mass spectrometry, and electron spectroscopy for chemical analysis. They have recently been described,[236, 237] and their analytic capabilities outlined for the nonspecialized reader. The limited number of suitable instruments yet in service and the early state of development of specific analytic procedures suggest that not too many applications to platinum-metal analysis have been described. However, it is known that ion-scattering spectrometry has been successfully used to determine platinum metals in supported catalysts where the metal is on the surface of a carrier of greatly different atomic mass.

ELECTRON-PROBE MICROANALYSIS[20, 239]

Electron-probe microanalysis depends on excitation of the characteristic x-ray spectra of elements by a narrowly focused beam of electrons directed to strike a small part of a sample to be analyzed. The emitted x-ray spectrum is measured by an x-ray spectrometer; qualitative information concerning the elements present in the sample is given by the wavelengths emitted. Quantitative analysis is achieved by measuring the intensity of the emission and making various corrections.

There have been some remarkably successful applications of this method to the analysis and characterization of new minerals of the platinum metals (among others), which are referred to in Chapter 2.

Neutron-Activation Analysis

Activation techniques are probably unequaled for the purpose of determining submicrogram traces of most metals. With a flux of 5×10^{-11} neutrons/cm²-s, neutron activation is at least one order of magnitude more sensitive for most platinum metals than the best of the spectrophotometric methods.

The sensitivities listed in Table 5-2 are predicated on irradiation of a sample for 1 month at a neutron flux of $10^{-2}/cm^2$-s, followed by a 2-h decay during which radiochemical purification with a quantitative yield is carried out.

Activation procedures have been applied to analysis of platinum catalysts. A 2-h irradiation at a relatively low neutron flux of $10^8/cm^2$-s was sufficient to detect platinum at 17 ppm and ruthenium at 38 ppm, without the necessity of any chemical separation steps. The other platinum metals should be as readily analyzed.

A complete list of isotopes, natural and artificial, of the platinum-group metals is given in Walsh and Hausman.[462]

Electrochemical Methods

POLAROGRAPHY

Polarographic behavior of the platinum metals has been extensively studied. Platinum may be determined in the presence of palladium at 0.02-0.2 μg/ml by the catalytic wave in 2 M hydrochloric acid. Palladium at 20-60 μg/ml may be determined in an EDTA solution at a pH of 5-7 without interference from the other platinum metals. It is possible to determine 0.5-25 μg of rhodium in 20 ml of a sodium chloride-hydrochloric acid solution by oscillographic polarography without interference from platinum, palladium, or gold. The polarographic reduction wave of iridium(IV) to iridium(III) in hydrochloric acid can be used for 1-450 ng/ml in the presence of rhodium and palladium. Osmium and ruthenium can be simultaneously determined by oscillographic polarography in sodium chloride-hydrochloric acid solutions.

TABLE 5-2 Sensitivity of Neutron-Activation Analysis for Platinum-Group Metals and Gold

Radioisotope Produced (half-life)	Estimated Sensitivity, g
Gold-198 (2.7 d)	5×10^{-12}
Ruthenium-105 (4.5 h)	1×10^{-9}
Osmium-193 (31 h)	1×10^{-9}
Rhodium-104 (4.3 min 44 s)	5×10^{-3}
Iridium-194 (19 h)	1×10^{-11}
Platinum-197,199 (18 h)	1×10^{-9}
Palladium-109 (13.5 h)	1×10^{-10}

Analysis and Determination

COULOMBMETRY

Controlled-potential coulombmetry has been used for accurate determination of microgram, as well as milligram, quantities of platinum, palladium, rhodium, iridium, and ruthenium. Coulombmetric generation of titrants, as well as direct reduction of the metal ion, has been effectively used.

AMPEROMETRY

Amperometric titrations are based on the use of polarographic waves as indicators of the consumption of the electroactive species by its reaction with a suitable titrant. Thus, 0.5–2.1 mg of palladium can be titrated amperometrically as it is oxidized from oxidation state +2 to +4 by standardized hypochlorite solution. Iridium and ruthenium can be titrated in the same way on reduction by ascorbic acid or hydroquinone as titrant.

Gravimetric Methods

Platinum may be precipitated quantitatively from hydrochloric acid solutions in the form of ammonium hexachloroplatinates, $(NH_4)_2PtCl_6$; very few base metals interfere, but iridium, rhodium, or palladium may interfere somewhat, and reprecipitation may be necessary (see Chapter 2).

Dimethylglyoxime, $C_4H_6(NOH)_2$, is the most important precipitant for palladium, which, in acid solution, is precipitated as $Pd(C_4H_7N_2O_2)_2$ essentially without interference from any of the other platinum metals or base metals.

No fully acceptable gravimetric methods are available for the other platinum metals, but procedures are available for use in restricted applications.[37] A classic case is the precipitation of ruthenium, after distillation and absorption of its tetroxide, by thionalide (α-mercapto-N,2-naphthylacetamide), followed by controlled ignition to the metal.

Volumetric Methods

Palladium may be titrated by potassium iodide in the presence of up to a 2-fold excess of platinum and a 200-fold excess of copper(II); the end point is found potentiometrically. Vanadium(II) may also be used as a titrant for palladium. An indirect method for the same metal, with which platinum does not interfere, involves the addition of an excess of EDTA and back-titration with zinc(II).

Platinum(IV) may be titrated with iron(II) sulfate containing triethanolamine in an alkaline medium; under these conditions, iridium(IV) is reduced to iridium(III), and palladium(II) is reduced to metal. Copper(I) chloride may be used to titrate platinum in the presence of palladium. Ruthenium(IV) may be potentiometrically titrated to ruthenium(VIII) by lead tetraacetate; alternatively, reduction of ruthenium(IV) to ruthenium (III) by iron(II) has also been used.

6

Toxicology and Pharmacology

The toxicology and pharmacology of the platinum-group metals include the study not only of the six metals themselves, but also of their compounds. The recent introduction of catalytic converters using platinum, palladium, and rhodium to control motor-vehicle exhaust emission makes it possible for quantities of the metals and their compounds to enter the environment (R. F. Hill, personal communication; and Brubaker et al.[66]). Radioruthenium can also enter the environment through the waste effluent from nuclear-fuel reprocessing plants, where it is one of the most abundant fission nuclides released. The exact amounts and chemical forms that may be lost are not known, but it is imperative to consider carefully the toxic and pharmacologic properties of these materials, if intelligent recommendations about their use are to be made.

Other potentially significant uses of platinum-metal compounds are their application as antitumor chemotherapeutic agents, several of which are undergoing limited testing.

The purpose of this chapter is to assess the physiologic activity of these materials, inasmuch as the human population may ultimately be exposed to them through the skin, by inhalation, and orally.

ACUTE TOXICITY OF PLATINUM AND PALLADIUM COMPOUNDS

Hofmeister[204] was one of the first to emphasize that the toxicity of platinum and its salts and complexes varies. The compounds tested were ammonium salts containing divalent and tetravalent platinum with various numbers of ammine ligands. He concluded that an increase in the valence of the noble-metal component of a complex salt was associated with a stronger "curare-like" action of the salt. The acute toxicity of a number of platinum coordination complexes has been investigated.[373] There is general agreement about the lethality of specific compounds, but the acute toxicities of the various compounds vary over a wide range, as indicated by the $LD_{50}s$* of both soluble and insoluble complexes. Studies of the toxicology of the antitumor agent cis-dichlorodiammineplatinum(II), cis-[Pt(NH$_3$)$_2$Cl$_2$], have been conducted on mice and rats and have been summarized by Rosenberg.[373] The important histologic changes observed were denudation of intestinal epithelium, bone marrow depression, thymic and splenic atrophy, and acute nephrosis. Studies in dogs and monkeys revealed similar toxic effects, with damage to the renal tubules prominent, especially at high doses. Other important effects were damage to the bone marrow and gastrointestinal epithelium.

The acute LD_{40} of platinous chloride, $PtCl_2$, administered as a single dose to outbred albino rats and followed for 2 weeks is approximately 26 mg/kg of body weight.

The LD_{50} of palladium dichloride, $PdCl_2$, has been determined after intravenous, intraperitoneal, oral, and intratracheal administration for several species. Some of the data for rats and rabbits are presented in Table 6–1. The greatest toxicity was seen after intravenous or intratracheal administration, and the least after oral administration. To examine the effect of the chemical nature of the salt on toxicity, values for the LD_{50} of palladium chloride, K_2PdCl_4, and $(NH_3)_2PdCl_4$ were determined after intravenous administration. The $LD_{50}s$, expressed in micromoles of palladium per kilogram of body weight, were very similar for these compounds. Results of preliminary experiments indicated that palladium chloride or palladium sulfate, $PdSO_4$, when injected intravenously, acts as a nonspecific cardiac irritant, as well as a peripheral vasoconstrictor. Inasmuch as the chloride salt strongly dissociates in solution, the palladium itself may be the irritant.

In a study[205] of the toxicity of various lead, manganese, platinum, ruthenium, and palladium salts after oral administration, the toxicities

---

*LD_{50} is the dose at which 50% of the treated animals die.

Toxicology and Pharmacology

TABLE 6-1 Approximate LD_{50}s of Palladium Chloride[a]

Animal	Approximate LD_{50}, mg/kg of body weight	Route of Administration[b]
Rat	5	Intravenous
Rat	70	Intraperitoneal
Rat	200	Oral
Rat	6	Intratracheal
Rabbit	5	Intravenous

[a]Data from Moore et al.[282]
[b]Clonic and tonic convulsions were noted after intravenous administration. The manner of death after administration by the other routes was not mentioned.

of the salts were, in decreasing order: $PtCl_4$, $Pt(SO_4)_2 \cdot 4H_2O$ > $PdCl_2 \cdot 2H_2O$, $RuCl_3$ > $MnCl_2 \cdot 4H_2O$, $PdSO_4$, $PbCl_2$, $PtCl_2$ > PtO_2 > MnO_2, PdO. Thus, on oral administration, the two soluble tetravalent platinum salts were the most toxic, and the highly insoluble salts, including the oxides of platinum and palladium, were the least toxic.

Palladium chloride has a slight diuretic effect in rats, and lethal intravenous doses in rabbits cause hemolysis, albuminuria, and diuresis. The heart, kidney, liver, bone marrow, and blood cells are the sites of tissue damage.[271] The dermal irritancies of 11 platinum and palladium compounds were evaluated with albino male rabbits. The procedures and evaluation criteria were adapted from those in use by the National Institute for Occupational Safety and Health. Four materials were evaluated as unsafe for contact with intact or abraded skin, as judged by severity of response: $(C_3H_5PdCl)_2$, $(NH_4)_2PdCl_4$, $(NH_4)_2PdCl_6$, and $PtCl_4$. One was evaluated as safe for intact, but not abraded, skin: K_2PdCl_6. Two were found safe for intact skin, but for abraded skin only if protected: K_2PdCl_4 and $PdCl_2$. The remainder were considered safe for intact or abraded skin (irritancy grade, less than 1 on a scale of 4): $Pd(NH_3)_2Cl_2$, PdO, PtO_2, and $PtCl_2$.[77]

Two platinum compounds (PtO and $PtCl_2$) and two palladium compounds (PdO and $PdCl_2$) were examined for ocular irritancy in rabbits.[224] In each rabbit, 10 mg of the test material was deposited on the surface of the right eye, and the left eye was used as a control; six rabbits were used per compound. The rabbits were examined for ocular inflammation 24, 48, and 72 h after application. At these doses, neither of the platinum compounds was irritating, and no reaction was noted with PdO. All six animals that received $PdCl_2$ showed corrosive conjunctival lesions and severe inflammation of the cornea and anterior chamber of the eye. These effects were seen at 24 h and persisted at 48 and 72 h.

Dermal and ocular irritancies of compounds of metals other than platinum and palladium can be obtained from other sources.[458]

REACTIONS WITH ENZYMES, NUCLEIC ACIDS, BACTERIA, AND VIRUSES

The interactions of platinum-group metals and their complexes with biopolymers are still poorly understood, despite at least two strong motivations for studying them—the use of heavy-metal derivatives for isomorphous replacement in proteins for x-ray crystallographic determinations of structures, and the antitumor activity of many of these complexes.

Platinum–Protein Interactions in X-ray Crystallography

Crystalline proteins steeped in various mother liquors produce clear x-ray diffraction patterns. To reduce these patterns to atomic structures, the phases must be determined. To accomplish this, a heavy atom could be attached to specific sites of the protein if it does not distort the diffraction patterns (isomorphous derivatives). A solution of the heavy-atom compound is added to the mother liquor, and the crystals are allowed to interact for long periods to saturate possible binding sites. Among the most successfully used heavy atoms are those of the platinum group, particularly the $[PtCl_4]^{2-}$ ion. As expected, the evidence (reviewed recently by Petsko et al.[324]) suggests that the $[PtCl_4]^{2-}$ ions selectively attack specific protein sites, such as disulfide bonds, methionines, histidines, NH_2 terminal groups, and sulfydryls, with a rare few proteins showing binding at asparagine (hexokinase 2) and arginine (lysozyme chloride). The major binding sites are listed in Table 6–2. These are consistent with expectations that the $[PtCl_4]^{2-}$ ions, as well as other platinum complexes, preferentially bind to nitrogen and sulfur. The bonds are mainly covalent, with S_N2 substitution at the chloride positions. Stereochemistry is a major determinant of the specific sites of interaction, because of the rigid square-planar structure of the platinum(IV) complexes. The chemistry of the complexes in the mother liquor is not well understood. Dickerson et al.[123] have suggested the formation of an intermediate platinum(IV) complex with the protein ligands attached at the fifth and sixth coordination sites of the resulting octahedral complex. Petsko et al.[324] argued that this is less likely to occur in the nonoxidizing environment of the protein mother

TABLE 6-2 Survey of $[PtCl_4]^{2-}$ Binding to Protein Crystals[a]

Protein	Sites of $[PtCl_4]^{2-}$ Binding[b]
Carboxypeptidase A	S–S 138–161; Met 103; His 303; N–Term.
α-Chymotrypsin	S–S 1–127; Met 192; N–Term.
γ-Chymotrypsin	S–S 1–127; Met 192;
PMS-δ-Chymotrypsin	Met 192; S–S 1–127; His 57
Concanavalin A	Met 130
Cytochrome	
Horse-oxidized	Met 65; His 33
Tuna-reduced	Met 65
Erythrocrurorin	Met 131; His 94; His 111
Hexokinase 2	As 1
Lysozyme chloride	Arg 14
Prealbumin	–SH
Ribonuclease S	Met 29
Subtilisin BPN	Met 50; His 64
Thermolysin	His 250; His 216

[a]Data from Petsko et al.[324]
[b]S–S, disulfide link; Met, methionine; His, histidine; –SH, sulfhydryl; Arg, arginine; As, asparagine.

liquor and favored a scheme whereby some of the chlorides are replaced by ammonia or $(PO_4)^{3-}$ of the mother liquor to produce platinumdiammine complexes, which then react with the protein sites.

Robertus et al.[369] recently reported a structural determination of yeast phenylalanine tRNA to a 3-Å resolution using trans-dichlorodiammineplatinum, trans-$Pt(NH_3)_2Cl_2$, and an osmium–adenosine triphosphate complex as two of the isomorphous derivatives. The trans-dichlorodiammineplatinum has a major binding site on the anticodon oligonucleotide sequence Gm–A–A–Y–A–ψ_p, one of the most significant regions of the tRNA molecule.[358] Interestingly, the cis-dichlorodiammineplatinum did not show a similar interaction at this site, nor did it yield a satisfactory isomorphous derivative. This emphasizes the major importance of stereochemistry in determining the reactions of these complexes with biopolymers.

Platinum–Protein Interactions in Enzyme Inhibition

$PdCl_2$ is known to bind to various proteins—such as carboxypeptidase, casein, papain, and silk fibroin—but the binding sites are not known. Enzyme inactivations by $PdCl_2$ were reported by Spikes and Hodgson.[422] Only trypsin and chymotrypsin were rapidly inactivated at a pH of 4.2, and even these were much less affected at a pH of 8.9. The authors speculated that the inactivation may occur through free sulfhydryl or cystine groups in trypsin and through cystine groups in chymotrypsin. Catalase, lysozyme, peroxidase, and ribonuclease were not inhibited at all.

Inhibition of the two enzymes, leucine aminopeptidase and malate dehydrogenase, by a variety of platinum complexes has been studied by a group at Auburn University for some years. Leucine aminopeptidase is inactivated by $PtBr_4$, $Pt(En)Br_2$, and $Pt(Dien)Br^+$, with the rates and final percentage of inhibition decreasing in sequence. These studies were done at a pH of 8.0. The authors[173] suggested that at least two halide-leaving groups are required for high inactivation rates. Additional studies showed that the monoaquo complex $PtBr_3(H_2O)^-$ inhibits leucine aminopeptidase and malate dehydrogenase more rapidly than $PtBr_4^{2+}$, which is consistent with the suggestion by the authors and others that the slow aquation is the rate-limiting step in reactions with biopolymers:

$$Pt(X)halide + H_2O \xrightarrow{slow} Pt(X)H_2O + halide$$

and

$$Pt(X)H_2O + S \xrightarrow{fast} Pt(X)S + H_2O,$$

where S is some biopolymer group bound to the platinum.[272]

Although K_2PtCl_4 and Rb_2PtBr_4 produce the same equilibrium inhibition (after 24 h of incubation) of malate dehydrogenase, the bromide complex approaches the equilibrium approximately 5 times faster than the chloride complex. This is consistent with the ratio of the known rates of aquation of these two ligands and supports the hypothesis that aquation is the rate-limiting step in substitution reactions with protein ligands.[149]

As a general rule, neutral platinum chloroammine complexes that are active antitumor agents are in the *cis* configuration; the corresponding *trans* isomers are inactive. Thus, stereospecificity is of great biologic significance. This provides a powerful tool for elucidating the molecular mechanism of action. It is assumed (but not proved) that the *cis* isomers attack significant sites by a bidentate binding (chelation) at neighboring sites, whereas the *trans* isomers attack by monodentate substitution or, if bidentate, at obviously different spatial sites. Inasmuch as both the *cis*- and *trans*-dichlorodiammineplatinum isomers inhibit some enzymes (malate dehydrogenase and liver alcohol dehydrogenase) with the same association inhibition constant, it is proposed that the inhibition of these is by a monodentate substitution. For other enzymes (yeast alcohol dehydrogenase and lactate dehydrogenase), the association inhibition constant (or equilibrium association constant), K_e, is approximately 20 times higher for the *trans* isomer than for the *cis* isomer. It is suggested that a bidentate chelation over a wider gap is the essential interaction of binding and inhibition.[150] The nature of the binding ligands of these proteins is still undetermined. Table 6–3 shows some equilibrium association constants for a number of different platinum complexes with malate dehydrogenase.

Platinum-Complex Interactions with Amino Acids

Amino acids have at least four groups that react readily with platinum and other platinum-group metals: $-NH_2$, $-CO_2$, $-SCH_3$, and the imidazole ring nitrogens of histidine. Extensive work by the Russian school, and particularly the Volsteyn laboratory, has elaborated the kinetics and thermodynamics of reactions of $[PtCl_4]^{2-}$ 1 and of *cis*- and *trans*-dichlorodiammineplatinum with these amino acid groups. These have been reviewed in some detail by Thomson et al.,[439] McAuliffe and Murray,[270] and D. R. Williams.[481]

TABLE 6–3 Equilibrium Association Constants (K_e) for Platinum Complexes and Malate Dehydrogenase[a]

Complex	$K_e[M]^{-1} \times 10^2$
$PtCl_4^{2-}$	8,700
$Pt(NH_3)C_3^-$	290
cis-$Pt(NH_3)_2Cl_2$	3.0
trans-$Pt(NH_3)_2Cl_2$	3.2
$Pt(NH_3)_3Cl^+$	No observed enzyme inhibition
$Pt(NH_3)_4^{2+}$	No observed enzyme inhibition
$PtCl_6^{2-}$	5,200
$PtBr_6^{2-}$	4,300
$PtBr_4^{2-}$	9,200

[a]$K_e = \dfrac{[E \cdot I]}{[E] \cdot [I]}$, where [E] = enzyme concentration, [I] = free platinum complex concentration, and [E · I] = enzyme–platinum complex concentration. Data from Friedman and Teggins.[150]

A variety of reaction products are formed, depending on the ratio of the concentrations of the reactants, the pH, and the isomeric structure of the platinum complex. For example, $[PtCl_4]^{2-}$ reacts with glycine to give the bis-(*cis*)-glycinatoplatinum(II) complex and the *trans* isomer, in a 3:1 ratio. In these cases, both $-NH_2$ and $-CO_2^-$ of each glycine are bound to the platinum. At a higher pH and with excess glycine, the Pt(glycine)$_4$ complex results. Thomson *et al.*[439] pointed out that, as a general rule, at neutral pH the preferred form is the closed-ring structure, but at higher and lower pH the rings are open. In the platinum chloroammines, only the chloride is substituted, and the *trans* labilizing influences of the incoming amino acid groups are very slight. However, in the reaction of the $-SCH_3$ group of methionine (in excess) with *cis*-dichlorodiammineplatinum, the *trans* labilizing effect of the Pt-S bond is strong, and the ammonia ligand that is *trans* to the $-SCH_3$ is replaced by the $-NH_2$ of the amino acid, to form a bidentate closed ring on one side of the molecule and an open monodentate ligand opposite it.

McAuliffe and Murray[270] studied the reactions of palladium(II) with the sulfur-containing amino acids—in particular, methionine. X-ray crystallography has confirmed the structure of the resulting complex as the dimer connected by hydrogen bonds from the

$$\left[\begin{array}{c} Cl \\ \diagdown \\ Pd \\ \diagup \\ Cl \end{array} \begin{array}{c} NH_2 - CH \\ \diagup \quad \diagdown \\ \quad \quad CH_2 \\ \diagdown \quad \diagup \\ S - CH_2 \\ | \\ CH_3 \end{array} \begin{array}{c} COOH \\ \end{array} \right]_2$$

carboxylic acid. With *S*-methylcysteine, the results were Pd$_2$(Cyst)$_3$Cl · 2H$_2$O and Pd(Cyst)Cl.

Platinum–Nucleic Acid Interactions

The fact that some platinum coordination complexes are effective anticancer agents in animals and man has provoked a series of studies of these complexes in many laboratories on their molecular mechanisms of action. The results of such studies, which strongly implicate DNA as the target cell receptor, have been reviewed by Rosenberg,[373] Roberts,[365] Thomson,[438] and Robins.[370]

Toxicology and Pharmacology

PLATINUM-NUCLEIC ACID INTERACTIONS *IN VIVO*

cis-Dichlorodiammineplatinum(II) and various analogues cause tumors in animals to regress. Several of these drugs are now in clinical trial, and they have been studied widely for their actions in cells. Apparently, the effective drugs' ligand structures remain intact after injection into animals, owing to the high extracellular chloride concentration (0.112 M), which represses chloride dissociation. Thus, the drugs are passively transported through cellular membranes. However, the low intracellular chloride concentration (0.004 M) allows dissociation of the chloride ligands and their sequential replacement with aquo ligands; the resulting monoaquo or diaquo species react with specific cell receptors to produce biologic actions.

The effects of the drugs on the synthesis of cellular biopolymers (DNA, RNA, and proteins) have been studied by Harder and Rosenberg[184] and by Howle and Gale[212] in both tissue culture and mice. The generally consistent results are:

• At low doses (equivalent to the amount found in tumor tissue, 5 μg/g of tissue after a therapeutic dose), synthesis of new DNA is selectively and persistently inhibited.

• RNA and protein synthesis is only moderately affected.

• The degree of inhibition of synthesis is dose-dependent, and the inhibition reaches a nadir about 4–6 h after removal of the drug pulse. This dose is not frankly cytotoxic; after 3–4 days, the resulting giant cells revert to apparently normal cells. At much higher doses, the drug *is* frankly cytotoxic, and all biosynthesis is inhibited. At the lower dose, synthesis of DNA precursors is not inhibited, nor is their transport through membranes affected. The inhibition of DNA synthesis cannot be reversed by extensive dialysis or washing of the cells; this indicates tight binding of the platinum to the cell receptor.

All these results suggest that the primary lesion in the cell caused by the drug is an attack on the DNA of the cell. Table 6–4, derived from Harder,[183] shows the distribution of the platinum molecules among the various cellular biopolymers when *cis*- and *trans*-dichlorodiammineplatinum(II) is given to the cells at doses that would reduce the surviving cells to 0.37 of the initial number. In conclusion, the reactions of the platinum complexes with DNA lead to biologic activity.

PLATINUM-DNA REACTIONS *IN VITRO*

Horáček and Drobník[209] followed the interactions of *cis*-dichlorodiammineplatinum(II) and DNA with ultraviolet difference spectro-

TABLE 6-4   Yield of cis- and trans-Dichlorodiammineplatinum(II) Bound to HeLa Cell Macromolecules of Average Molecular Weight at the Mean Lethal Dose[a]

		No. Molecules Bound When Surviving Fraction Reduced to 0.37 of Original	
Molecule	Approximate Molecular Weight	cis-Dichlorodiammine-platinum(II) Molecules	trans-Dichlorodiammine-platinum(II) Molecules
DNA	1×10^9	22/DNA molecule	125/DNA molecule
mRNA	4×10^6	1/8 mRNA molecules	2.5/mRNA molecule
rRNA	$0.5-1 \times 10^6$	1/30 rRNA molecules	1/2 rRNA molecules
tRNA	2.5×10^4	1/1,500 tRNA molecules	1/70 tRNA molecules
Protein	1×10^5	1/1,500 protein molecules	—

[a] Derived from Harder.[183]

photometry. They reported a shift in the absorption maximum of DNA from 259 to 264 nm. At a platinum:phosphorus ratio of 1:1, there was also a marked hyperchromicity. At lower platinum:phosphorus ratios, they found a significant amount of renaturation after a heating and cooling cycle, and they suggested that this was due to cross-linking of the double-stranded DNA. Chemicals that are known to stabilize the DNA structures (e.g., putrescine, spermidine, histones, and Mg^{2+}, SO_4^{2-}, and NO_3^- ions) did not change the reaction rates or the final equilibrium values. High chloride concentrations did retard the reaction rates, probably by slowing the hydrolysis reaction of the platinum drug, which is assumed to be the rate-limiting step. The measured activation energy (ΔH) for the reaction was $18,000 \pm 630$ cal/mol.

The two cis chloride-leaving groups of cis-dichlorodiammine-platinum(II) are 3.3 Å apart. Therefore, it appears unlikely that the DNA is cross-linked at the N_7 position of guanosine in the two opposing strands, as has been suggested for the attack of bifunctional alkylating agents on DNA. Roberts and Pascoe[366] studied the cross-linking ability of both cis- and trans-dichlorodiammineplatinum(II) in DNA, both in vivo and in vitro, with the BUdR labeling technique. They concluded that the cis and trans isomers are both capable of cross-linking DNA, but a much higher dose of the trans isomer is required to produce the same amount of cross-linking as the cis isomer. The evidence of cross-linking of DNA has been verified with other techniques in other laboratories.

Thomson[438] has suggested that, in addition to forming interstrand cross-links, the platinum drug may cause intrastrand cross-links involving the 6-NH$_2$ groups of adenine in an ApT sequence, the 2-NH$_2$ groups of guanine in the narrow groove, and the 6-NH$_2$ groups of cytosine in a CpG sequence in a deep groove.

Because only a small number of the platinum molecules bound to DNA produce cross-linking, there must be other modes of binding. Stone et al.[431] have shown that the platinum binds predominantly in GC-rich regions of the DNA, and, indeed, they can assay the GC content of any DNA after reaction with the platinum drug by using density-gradient centrifugation. The platinum drug does not react well with histones, and chromatin appears to bind somewhat less of the drug than comparable amounts of pure DNA. Mansy[268] studied the reaction of cis-dichlorodiammineplatinum(II) and DNA with laser Raman light-scattering. The results show no evidence of interaction of the platinum complex with either the phosphate or sugar moieties of the DNA. Robine,[371] however, in his studies of the rate of reaction of dichloroethylenediamineplatinum(II) with DNA and other precursors, found that the presence of the phosphate group increased the reaction rate. The lack of evidence of phosphate binding from the laser Raman studies could be explained by assuming that the phosphate–platinum interaction is a short-lived intermediate in the final reaction of the drug with the DNA bases.

Bacterial and Viral Effects of Platinum-Group-Metal Complexes

Metal complexes have a long history of clinical use in bacterial infections. However, the involvement of platinum-group metals in chemotherapy started in 1953 with Shulman and Dwyer[409] and co-workers. They showed that the relatively inert chelate complexes of trisphenanthroline ruthenium are active bacteriostatic and bactericidal agents against gram-positive microorganisms, but are less active against gram-negative microorganisms. Substituted derivatives are also somewhat active against the Landshutz ascites tumor in mice. Unfortunately, these types of charged complexes exhibit intense neuromuscular toxicity ("curare-like" behavior) and can be used only topically; however, clinical tests did indicate their usefulness in the treatment of dermatosis, dermatomycosis, and monilial nail-fold infections.[409]

Rosenberg and his co-workers[376] first called attention to the effects of simple platinum-group-metal complexes on microorganisms in 1965. In recent years, many investigators have studied these systems, and a multiplicity of different effects have emerged.

BACTERICIDAL EFFECTS

For the simpler complexes—such as $(NH_4)_2PtCl_4$ and the corresponding salts of iridium, osmium, and palladium—the dominant effect on microorganisms is bactericidal.[374] As a general rule, charged species in solution are bactericidal, and not bacteriostatic, as shown by viable-cell counts. The effective concentrations of the drugs vary from 1 μg/ml for $(NH_4)_2PtCl_6$ to 40 μg/ml for K_2RuCl_6. The distribution of a bactericidal charged metal complex among the various chemical components of the bacterial cells is shown in Table 6-5; the data were obtained by use of radioactive platinum-191 labels. Although the great preponderance (95%) of this complex is bound to cytoplasmic protein, as would be expected from the evidence presented above, no conclusion can be reached on the mechanism of the bactericidal action, other than that it is unlikely to be a direct attack on cellular DNA.

The neutral complexes generally force filamentation, but they are bactericidal at higher concentrations (> 38 μM). Shimazu and Rosenberg[405] showed that mutant *Escherichia coli* with deficient DNA repair mechanism (B_{s-1}, B_{s-2}) are more sensitive to the drug than normal *E. coli* B or an ultraviolet-resistant strain (B/r); this implicates an interaction with bacterial DNA as the lethal lesion for the neutral complex. There is some evidence that microbial resistance to the bactericidal activity of these drugs does not develop. Sensitivity to these drugs has not been examined in a wide array of bacteria. These subjects require much further work.

The bacteriostatic activity of rhodium complexes has been studied by Bromfield and co-workers.[60] The most active complexes are *trans*-[RhL$_4$X$_2$]Y, in which L is a substituted pyrimidine, X is a halide, and Y

TABLE 6-5 Average Distribution of Platinum-191[a] among Cellular Components of *Escherichia coli*[b]

	Percentage of Radioactivity	
Component	Double-Negative Species	Neutral Species
Metabolic intermediates	1	19
Lipids	3	6
Nucleic acids	1	30
Cytoplasmic protein	95	45

[a]Double-negative and neutral species of ammonium hexachloroplatinate.
[b]Derived from Renshaw and Thomson.[356]

is Cl, Br, NO$_3$, or ClO$_4$. These are active against gram-positive bacteria at concentrations of 0.1–5 μg/ml, but require concentrations 100 times larger for activity against gram-negative bacteria. Cobalt and iridium analogues were tested, but are not active. The authors suggested that cobalt complexes are too labile, whereas iridium complexes are too stable. Rhodium complexes are of intermediate lability and are therefore able to pass intact to the site of action in the cell (unknown) and react with the target site.

FILAMENTATION EFFECTS

Some platinum-group-metal complexes, when present in dilute concentrations in the growth media of microorganisms, selectively inhibit cell division without inhibiting growth, thus forcing the bacterial rods to form long filaments. Table 6–6 shows some typical results[374] of different metal complexes with *E. coli* as the target organism. Table 6–7 catalogs the effect of one complex, *cis*-dichlorodiammineplatinum(II), on a variety of gram-positive and gram-negative organisms. The latter group are by far the more sensitive. In addition, Bromfield et al.[60] reported that [Rh(pyridine)$_4$Cl$_2$]Cl and [Rh(4-ethylpyridine)$_4$Br$_2$]Br in-

TABLE 6–6 Effects of Group VIIIB Transition-Metal Compounds in Producing Elongation in *Escherichia coli*[a]

Metal Compound	Effective Concentration, μg/ml	Elongation[b] (estimated)
(NH$_4$)$_2$PtCl$_6$	1–20	100%, 25–100X
RhCl$_3$	30–100	75%, 5–25X
(NH$_4$)$_2$RhCl$_6$	20–30	75%, 5–20X
K$_2$RuCl$_6$	40–50	50%, 5–20X
RuK$_2$NOCl$_5$	40–100	50%, 5–10X
RuI$_2$	20–30	25%, 10–20X
RuNOCl$_3$	20–40	25%, 4–6X
RuBr$_2$	10–20	10%, 10–20X
RuNO(NO$_3$)$_3$	10–100	10%, 4–6X
K$_3$Rh(NO$_2$)$_6$	20–60	10%, 3–5X
(NH$_4$)$_3$RuCl$_6$	15–30	10%, 3–5X
UO$_2$(C$_2$H$_2$O$_2$)$_2$	25–75	5%, 3–5X
RuCl$_3$	—	—

[a]Derived from Rosenberg et al.[374]
[b]Growth is the estimate of the relative percentage of filamentous cells in a drop of culture fluid; the relative length of the filaments is expressed in comparison with a control of normal-size cells after 6 h of incubation in a synthetic medium.

TABLE 6-7 Elongation Effect of cis-Dichlorodiammineplatinum(II) on Bacteria in Nutrient Media[a]

Organism	Elongation[b]		
	5 μg/ml[c]	25 μg/ml[c]	75 μg/ml[c]
Escherichia coli B	50%, 10–25X	95%, 10–50X	Toxic
E. coli C	40%, 10–15X	70%, 10–25X	Toxic
E. coli K-12	25%, 5–10X	20%, 10–20X	Toxic
Aerobacter aerogenes	50%, 2–15X	Toxic	Toxic
Alcaligenes faecalis	25%, 2–10X	10%, 10–15X	Toxic
Proteus mirabilis	10%, 2–5X	25%, 5–10X	Toxic
Pseudomonas aeruginosa	25%, 5–10X	Toxic	Toxic
Klebsiella pneumoniae	50%, 2–10X	Toxic	Toxic
Serratia marcescens	20%, 2–5X	Toxic	Toxic
Bacillus cereus	Normal	Normal	25%, 2–10X
B. licheniformis	Normal	Normal	90%, 2–5X
B. megaterium	Normal	Normal	20%, 2–5X
B. subtilis	Normal	Normal	20%, 2–4X
Lactobacillus sp.	Normal	Toxic	Toxic
Clostridium butylicum	Normal	Normal	Toxic
Corynebacterium sp.	Normal	Normal	Toxic
Streptococcus lactis	Normal	Normal	Normal
S. faecalis	Normal	Normal	Normal
Staphylococcus aureus	Normal	Normal	Normal
Sarcina lutea	Normal	Normal	Normal
Neisseria catarrhalis	Normal	Normal	Normal

[a] Derived from Rosenberg et al.[374]
[b] Expressed in the same terms as those of Table 6-6; incubation time, 12–24 h.
[c] Concentration of platinum complex.

duce long filaments (20–500 times normal length) in 100% of the E. coli at concentrations of 5 μg/ml and 0.4 μg/ml, respectively.

Most of the biochemical studies of this effect involve cis-dichlorodiammineplatinum(II) and E. coli. Here, filamentation is a reversible effect; removing the bacterial filaments to a fresh medium free of the drug causes reversion to normal colonies of the bacteria. This appears to differ from filamentation caused by such agents as the nitrogen mustards, in which case filamentation is terminal. The synthesis of new DNA is not markedly impeded at the low concentrations of the drug that cause filamentation, and the bacterial DNA appears throughout the filament, either in clumps or as continuous, long, axial fibers. Filamentation is not reversed by pantoyl lactone, divalent cations, or temperature shock, which do cause reversal for most causative agents. As shown in Table 6-5, the neutral complex is distributed among the various classes of cellular components differently from the charged complex. About 30% is associated with nucleic acids, compared with 1% of the charged species of the drug. An indirect effect of the drug on bacterial respiration and reversible changes in the inducibility of β-galactosidase have also been reported.

Gillard, Harrison, and Mather (cited in Cleare[97]) tested a large number of rhodium complexes for their ability to inhibit cell division in *E. coli*, in an attempt to correlate it with some physicochemical property of the complexes. Most of the active complexes have polarographic half-wave reduction potentials between -190 and $+90$ mV, with respect to the standard hydrogen electrode. Inactive complexes were generally more negative. This may be related to a required reduction of rhodium(III) to rhodium(I) for activity. Steric factors and lipophilic character also seem to be correlated with activity.

LYTIC INDUCTION OF LYSOGENIC BACTERIA

A most interesting new bacteriologic effect of *cis*-dichlorodiammine-platinum(II), discovered by Reslova,[357] is that it induces bacteriophage virus from lysogenic strains of *E. coli*. This is consistent with earlier results with some organic antitumor agents; they are also potent depressors of latent viral information in lysogenized bacteria. The platinum drug produces detectable increases in phage-particle production, even when present in the bacterial growth medium at concentrations as low as $0.1\,\mu$M. This is well below the threshold for forcing filaments ($\simeq 1\,\mu$M) or for toxicity ($\simeq 30\,\mu$M). The number of induced (lysed) cells is increased over background at even lower concentrations ($\simeq 0.03\,\mu$M) of the drug. The platinum drug induces lysis in a wide variety of lysogenic strains and different bacteria. A positive correlation was found between the ability of a platinum complex to induce lysis and its antitumor activity.

In the direct induction of lysogenic bacteria, the inducer in the cell may directly or indirectly inactivate the phage repressor in the lysogenic cell. Another means of induction is even more indirect. In this case, the platinum drug is added to a nonlysogenic strain with a sex-specific F factor.[183(pp. 105-106)] This F factor (a DNA-containing replicon) is transferred by sexual conjugation to a lysogenic receptor cell, which then undergoes lysis. Because only DNA is transferred from the treated cell (F^+) to the recipient cell (F^-), it is very likely that the platinum-drug lesion on the DNA is responsible for induction.

VIRAL INACTIVATION *IN VITRO* AND *IN VIVO*

Viral particles incubated with various platinum complexes for different periods can be inactivated.[249,408] This is tested by transferring the treated virions to a suitable host organism, where their presence is manifested by a detectable pathologic effect. Bacteriophages T_{even}, T_3, T_7, R_{17}, and ΦX and SV 40, Rous sarcoma virus, influenza virus, Newcastle disease virus, and fowl pox virus have been shown to be inactivated by *in vitro* treatment with various platinum drugs. The most potent inactivator was the *cis*-diaquodiammineplatinum(II). Inactivation was rapid (minutes) and complete with this drug.

In vivo viral inactivation is measured by inoculating a suitable host with the live virus, treating the host later with the platinum drug, and following the course of pathologic events in the host.[183] It is of some significance that a delay between viral inoculation and drug treatment can occur without significant loss of activity. During this delay, the viral particles disappear into the cells (eclipse phase) and start replication. The results of one such study are shown in Table 6-8 for fowl pox virus in an embryonated egg host. The dose of the drug used was not frankly cytopathic to the host cells.

TABLE 6-8 Antiviral Activity of *cis*-Diaquodiammineplatinum(II) against Fowl Pox Virus *in Vivo*[a]

Delay between Virus and Drug, h	Reduction of Pox Lesion, %
0	97.2
2	93.3
4	98.2
6	78.2
8	91.3
10	66.7
24	23.7

[a]Derived from Harder.[183] Fifteen FPV particles suspended in 0.1 ml were inoculated at time zero into an artificial air cell previously prepared on the chorioallantoic membrane (CAM) of each embryonating white leghorn egg. At various intervals, 0 or 3.9 μg of *cis*-diaquodiammineplatinum(II) in 0.2 ml of water was inoculated onto the CAM at the same site that had previously received the virus. On the 5th-6th day of incubation, the CAM was excised and washed, and the visible pox lesions in the membrane from treated and untreated eggs were counted.

Inactivation appears to require the aquation of the chloride-leaving groups of *cis*-dichlorodiammineplatinum(II), in agreement with other studies showing that the interaction of the drug with nucleic acid occurs via the intermediate diaquo form. It is therefore most likely that the viral inactivation is due to a reaction between the platinum drug and nucleic acid, but possible reactions with other biopolymers cannot be ignored. The kinetics of *in vitro* inactivation are first-order and reflect the kinetics of binding of the platinum drug to the virion. In the case of the SV 40 virus, the inactivation of infectivity and ability to induce V antigen occurs faster than the inactivation of the T antigen induction. The inactivated virions retain their immunogenic properties.

PHARMACOLOGIC DISPOSITION OF THE PLATINUM-GROUP METALS

The fate of platinum in rats has been studied by administering platinum-191 or platinum-191,193 as a chloride or as a sulfate intratracheally, orally, intramuscularly, and intravenously.[282,397]

The route of administration is important in determining the retention of platinum-191, as shown in Figure 6-1. Retention is greatest after intravenous administration, next greatest after intratracheal administration, and least after oral administration. In addition, retention after intratracheal administration is greater than that after inhalation (Figure 6-2).[282] In agreement with these data is the fact that most of the platinum-191 given orally is excreted in the feces, whereas platinum-191 given intravenously is excreted in both the urine and the feces, with urine containing a great quantity of it. Tissue distribution studies[397] 7 days after platinum-191,193 administration showed that the kidneys contain the greatest amount of radioactivity per gram, followed by the spleen and liver. Data derived from pregnant rats[282] given platinum-191 on the eighteenth day of gestation and sacrificed 24 h later revealed a small amount of transplacental passage of the isotope (0.01%/g of fetus), but there appears to be placental binding or accumulation.

The fate of palladium has also been studied in rats with $^{103}PdCl_2$ given by the intratracheal, inhalation, oral, and intravenous routes.[282] As with platinum, the amount of palladium-103 retained after a single dose depends on the route of administration. The retention declines rapidly after oral administration to about 0.4% of the initial dose after 72 h; less rapidly after intratracheal administration; and least rapidly after intravenous administration, with approximately 10% of the initial intravenous dose being retained within the body 76 days later. The amount of palladium-103 retained after intratracheal administration is

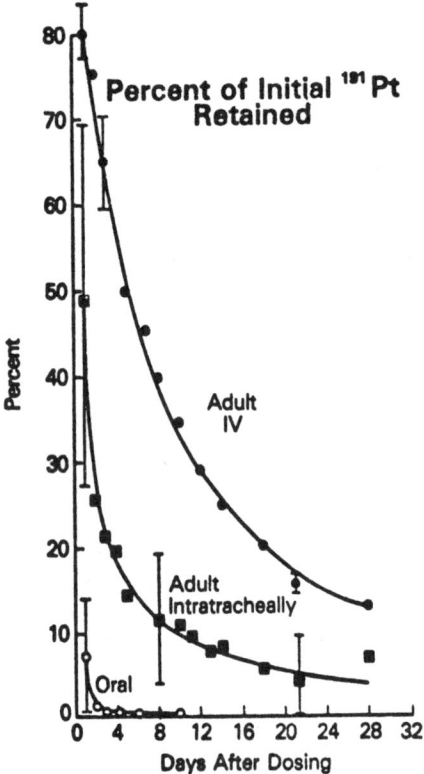

FIGURE 6-1 Whole-body retention of platinum-191 in adult rats after intravenous, intratracheal, and oral administration. Reprinted from Moore et al.[282]

higher than that after inhalation exposure. In addition, the retention time for palladium-103 declines rapidly in the suckling rat after oral administration, in a fashion similar to that in the adult; however, the amount absorbed and retained with time is significantly higher (Figure 6-3). It is interesting that, for a given route of administration, platinum and palladium appear to be excreted at about the same rate.

The fate of rhodium in rats was studied with carrier-free rhodium-105 given intravenously or orally.[126,397] Rhodium-105 was not found to a great extent in any of the tissues 96 h after oral administration, the kidneys having the highest amount—0.04% of the administered dose. However, after intravenous administration, 45% of the dose was still

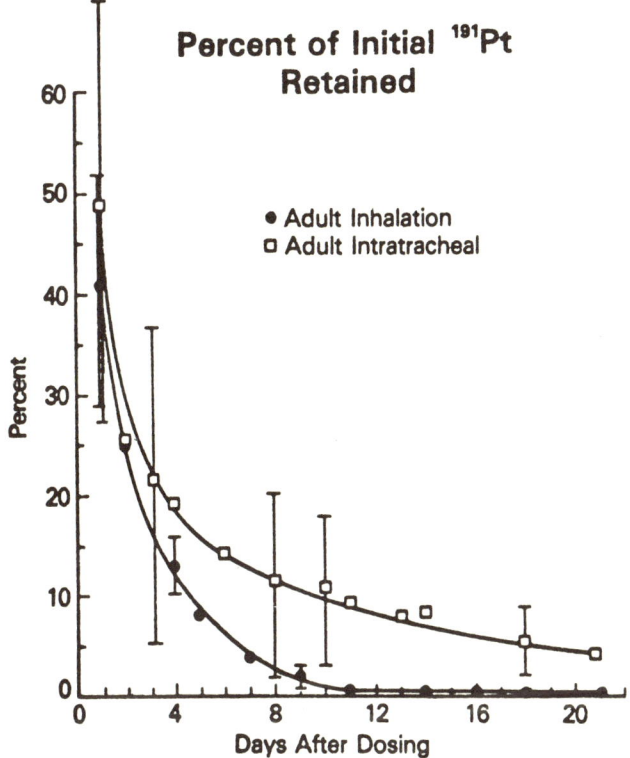

FIGURE 6-2 Whole-body retention of platinum-191 in adult rats after inhalation and intratracheal administration. Reprinted from Moore et al.[282]

retained by the rat, with the greatest amount being in the kidneys—1.5%/g of kidney. Prolonged application of rhodium chloride to rabbit skin results in degenerative changes in the kidneys and liver (this is also true for platinum, but not for palladium and ruthenium).[246]

The fate of carrier-free iridium-192 was measured in rats 2 h, 7 days, and 33 days after intravenous injection.[126,397] Two hours after injection, 18% of the iridium had been excreted in the urine, and the greatest tissue concentrations on a percent-per-gram basis were in the liver, kidneys, and lung. Seven days after injection, 38% had been excreted in the urine and 14% in the feces. The kidneys, liver, and spleen contained the highest amounts on a per-gram basis. Thirty-three days after intravenous injection, 44% had been excreted in the urine and 36% in the feces. Iridium-192 was still concentrated in the kidneys, liver, and

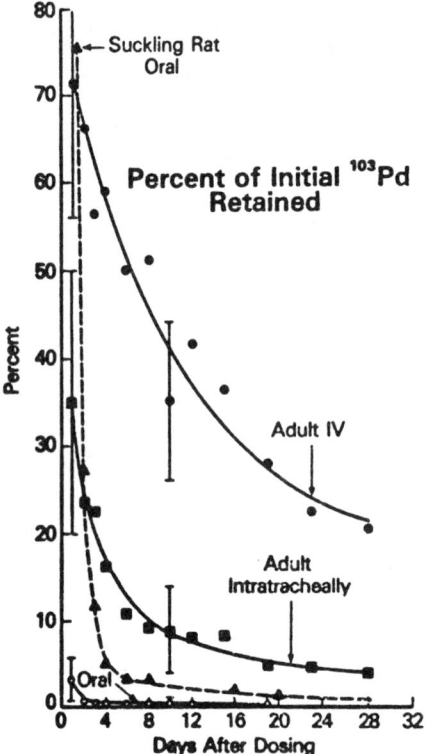

FIGURE 6-3 Whole-body retention of palladium-103 (given as PdCl$_2$) in adult rats after intravenous, intratracheal, and oral administration, and whole-body retention of palladium-103 in suckling rats after oral administration. Reprinted from Moore et al.[282]

spleen, both in percent per organ and in percent per gram. The fate of iridium-192 is similar to that of platinum. When iridium-192 is administered orally, only small amounts are found in the tissues and organs 7 days after injection, with 94% being excreted in the feces and 3.6% in the urine. The fraction absorbed from the digestive tract is estimated at 10%. Data on retention of iridium after inhalation have been reported by Casarett et al.[79]

The fate of carrier-free ruthenium has been evaluated by several investigators.[126,178,179,425] Studies have suggested that less than 0.5% is

Toxicology and Pharmacology

absorbed after an oral dose of ^{103}Ru^{4+} or ^{106}Ru^{4+} to rats, whereas chickens may absorb 3% of an intragastric dose of ruthenium-106.[180] Studies with five different ruthenium-106 compounds showed that the chemical form or compound in which radioruthenium is administered influences the total amount absorbed. Thus, for up to 2 weeks after intragastric administration to cats, 2.4% of RuO$_4$ is absorbed, but 12.3% of RuNO(NO$_3$)$_3$.[425] The distribution of ruthenium in the organs of cats is similar to that observed in other species. Thus, 5 days after administration, kidneys, bone, and testes retain the highest proportion of the ruthenium-106. Nelson et al.[299] have shown detailed autoradiographs of adult male and pregnant female mice at intervals of up to 32 days after intravenous injection. The metabolism of ruthenium in man has been examined by Yamagata et al.[493] and by Webber and Harvey.[467]

The retention of osmium after intravenous and intramuscular administration in rats has been reported by Durbin et al.[126] A few other studies of the toxicology of osmium have also been reported.[64,161,221,229,418,473]

In a study of the effects of various salts of platinum or palladium on the characteristics of the microsomal drug metabolizing enzyme system, with doses far exceeding anticipated environmental exposure, it was shown that the dietary administration of various salts of platinum or palladium for a week generally decreased or had no effect on the characteristics of drug metabolism by isolated microsomes. Four or more weeks after dietary administration of these salts, there was either no effect or, if anything, an increase in the measures used.[205] Considering the high doses used, one can only conclude that probable doses to which the human may be exposed would have little or no effect on the microsomal mixed-function oxidase system. Additional studies of the effects of platinum and palladium on different organ systems have also been reported.[144,283,464,478]

A recent study has attempted to establish baseline data for platinum concentration in human tissue before the widespread use of catalytic converters.[125] (A more detailed discussion appears in Chapter 8.) In addition to the tissue analyses, the following data are available for 92 of the 97 autopsy sets studied: age, race, sex, occupation, and cause of death. Clinical summaries, where available, were studied; these gave no indication that platinum tumor-suppressant therapy might have been a source of platinum exposure. Similarly, occupational information available did not indicate occupational exposure to platinum. Of a total of 1,313 samples obtained from the 97 subjects and analyzed for platinum, 62 contained detectable platinum; 45 subjects had detectable platinum in one or more tissue samples. Thus, although platinum was detected in only 5% of the samples analyzed, 46% of the subjects had detectable platinum in at least one type of tissue. Among the cases of detected platinum, the range of platinum concentrations was 0.003–1.46 μg/g (wet tissue), the mean was 0.16 μg/g, and the median was 0.067 μg/g.

The authors ranked the sites of platinum deposition in the subjects studied according to the frequency of detection of platinum in various tissue sample types. In order of decreasing frequency, they were: subcutaneous fat, kidney, pancreas, and liver. The presence of platinum in subcutaneous fat raises the question of transport, in that most platinum compounds are lipid-insoluble. Conversion to lipid-soluble compounds through methylation of platinum compounds is a possible explanation suggested by the authors.[125] However, the relatively small number of samples that actually contained platinum casts serious doubt on the statistical reliability of such analysis.

INHALATION TOXICITY OF PLATINUM-GROUP METALS AND COMPOUNDS

In view of the possible emission of small particles of platinum and palladium through attrition from automotive catalytic converters (see Chapter 8),[66,222] inhalation would be the most likely mode of human exposure. Indeed, inhalation of some platinum-containing salts or acids can provoke rather severe physiologic responses in people who have been "sensitized,"[148,218,328] as is explained in more detail in Chapter 7.

The most common current occupational exposure to noble metals is through inhalation of dusts by refinery workers involved in producing these materials.[218,276] It is not at all uncommon for an employee of such a plant to develop an asthmatic or dermatologic allergy that disappears when the person is removed from areas where the air is contaminated with these materials. It is thought that the complex platinum salts cause release of histamine that accounts for the asthma or hay-fever symptoms.[312–314,389] Apparently, the platinum metals themselves, even in a state of very fine subdivision, do not invoke significant physiologic responses when inhaled; the active species are some complex compounds of the metals. With respect to the two most common metals, compounds of platinum are by far more active allergens than those containing palladium. Recognition of this susceptibility has led the Occupational Safety and Health Administration to set a maximum of 2.0 μg/m³ (based on 24-h exposure) for exposure to airborne soluble platinum salts.[452] The standard is similar for osmium tetroxide,[452] and the maximal allowable exposure to soluble rhodium salts is set at 1.0 $\mu g/m^3$.[312]

CHRONIC EXPOSURE TO PLATINUM, PALLADIUM, AND RELATED COMPOUNDS

Animals

In experiments reported by Schroeder and co-workers,[394,395] toxic effects of small doses of scandium, hexavalent chromium, gallium, yttrium, indium, rhodium, and palladium on growth and survival in mice were evaluated. They raised 958 mice in an environment limited in metallic contamination and gave them metal at 5 ppm in drinking water from weaning until natural death. Body weight was measured every month for 6 months, at 12 months, and at 18 months. Compared with controls, the feeding of gallium was accompanied by significant but not marked suppression of weight at 14 of 16 measurements (eight in each sex); of scandium, at 10 measurements; of indium, at eight; of palladium, at seven; of rhodium, at six; of yttrium, at 12; and of hexavalent chromium, at eight. Survival of gallium-fed females was less than that of controls, whereas survival of palladium-fed males and yttrium-fed males and females was greater. Tumors were found at necropsy in 16.3% of one group of controls and in 27.4% of the scandium-fed, 26.0% of the gallium-fed, 13.0% of the indium-fed, 28.8% of the rhodium-fed, and 29.2% of the palladium-fed groups. Malignant tumors were increased in the rhodium and palladium groups, at a minimally significant level of confidence ($p < 0.05$), all but one tumor being malignant. In a second series, tumors were present in 26.8% of controls, 27.6% of the chromium-fed mice, and 33% of the yttrium-fed mice. All tumors in the latter two groups were malignant. Therefore, rhodium and palladium appear to exhibit slight carcinogenic activity in mice.[394]

A silver–palladium–gold dental alloy imbedded subcutaneously for 504 days caused tumor formation in seven of 14 rats.[152] Implants of a silver–palladium–gold–copper dental alloy (70.02, 24.70, 5.23, and 0.03%, respectively) in the oral submucous membranes of rabbits had only mild effects, but liver implants in rats temporarily constricted capillaries of the liver parenchyma, and testicular implants in rats caused seminiferous-tubule degeneration.[175]

Ridgway and Karnofsky[363] evaluated $PdCl_2$ for teratogenic effects in chicken eggs and found it to be nonteratogenic. The LD_{50} was greater than 20 mg/egg on the fourth day of incubation.

Studies evaluating platinum, palladium, and other members of this group for mutagenic, teratogenic, and long-term effects (other than the study reported above) are lacking, and investigation is urgently needed.

Osmium metal apparently is nontoxic, but the toxicity of the tetroxide in both animals and man is well known.[64] The volatile osmium tetroxide affects the eyes, causing a halo phenomenon, and also affects the lungs in both experimental animals and man.[64] Smith et al.[418,419] have reported more information on osmium.

Humans

The toxic and potentially toxic effects of platinum in humans are thought to involve the water-soluble platinum salts (potassium hexachloroplatinate, potassium tetrachloroplatinate, sodium chloroplatinate, and ammonium chloroplatinate), and not platinum itself.[1,148,218,364] However, Schwartz et al.[396] and Ledo-Dunipe[252] reported that dermatitis had resulted from contact with platinum oxides and chlorides, and the sensitization of skin to platinum during the process of soldering has been reported.

The subject of allergy to platinum and its compounds is covered more thoroughly in Chapter 7 and is introduced here because of its relevance. "Platinosis" has been defined as "the effects of soluble platinum salts on people exposed to them occupationally."[364] The condition is characterized by pronounced irritation of the nose and upper respiratory tract, with sneezing and coughing. In some cases, radiographic examination of the lungs has indicated a low-grade pulmonary fibrosis. The first symptoms, arising during exposure and persisting for about an hour after leaving work, are pronounced irritation of the nose and upper respiratory passages, with sneezing, running of the eyes, and coughing. Later, the "asthmatic syndrome"—with cough, tightness of the chest, wheezing, and shortness of breath— develops, becoming progressively worse with the length of employment.[64] Hunter et al.[218] reported that concentrations near the threshold limit value for platinum salts produce symptoms in some people. However, exposure to platinum-metal dust at much higher concentrations does *not* produce these symptoms among the industrial population.[364] The hypersensitivity response attributed to platinum is an important consideration in this matter. Evidently, a considerable proportion of the population responds to platinum salts as though they were allergens; once a person is sensitized, he never seems to become asymptomatic in a platinum atmosphere.[364] There are predisposing factors that should be considered regarding exposure of the general population to any amount of these materials. People with light complexion, light hair, blue eyes, and delicately textured skin appear to be considerably more susceptible than darker-skinned, brown-eyed people.[364] Long exposure of men to platinum and platinum compounds in refining has been associated with later formation of allergic reactions and dermatitis. However, even after 18-20 years of repeated exposure to threshold limit concentrations of platinum compounds, no increase in incidence of cancer has been observed.[222,364] Information on the mutagenesis of platinum and ruthenium complexes was given in a paper by Monti-Bragadin *et al.*[281]

A recent preliminary industrial-hygiene survey of workers who handle the catalytic material was conducted by the American Cyanamid Company (R. M. Cline and W. V. Andresen, personal communication). Workroom air concentration of platinum was near the threshold limit value (2 $\mu g/m^3$) for the water-soluble salts of platinum.[15] However, this material was considered to be the dust of the final catalyst product, not the water-soluble salts. All the workers had received annual physical examinations that demonstrated no signs of allergic reactions or other disease process, even after more than 10 years of exposure. A urinary-excretion study of these workers showed a significant increase in excretion of platinum at the end of a work cycle, compared with the beginning. In addition, the concentration of platinum was greater in workers at the beginning of a work cycle than in controls who presumably were not exposed to the dust.

Regarding other possible adverse health effects, including postulated aggravation of preexisting cardiorespiratory problems, Roberts[364] stated that "platinosis" does not appear to alter general health in any way other than to irritate the upper respiratory tract and to cause contact dermatitis.

Among the platinum-group elements, the allergic reaction is evidently peculiar to platinum. The only reported case[287] of contact dermatitis from palladium occurred in a research chemist who had been studying various precious metals for several months. His face, hands, and arms showed patches of eczema. The condition cleared up completely on avoidance of palladium exposure and treatment with betamethasone valerate ointment.

Palladium chloride has been used (ineffectively) to treat tuberculosis at a dosage of about 18 mg/day. Oral dosages up to 65 mg/day apparently produce no adverse effects.[271] Topical application of palladium chloride as a germicide does not cause any skin irritation.

Palladium hydroxide has been used to treat obesity. Colloidal palladium hydroxide injections (5-7 mg/day) reportedly caused a 19-kg weight loss in a 3-month period, with necrosis at the injection site.[271]

Palladium has been found in teeth that contained palladium-alloy fillings. Presumably, small amounts of palladium are solubilized by body fluids. Traces of palladium have also been found in human liver.[126]

Additional information on acute and chronic toxicity and pharmacologic properties of the platinum-group-metal compounds can be found elsewhere.[29,30,69,135,154,247,423,426,432,437,443,467,493]

7

Allergy to Platinum Compounds

The first report of undue reactions to platinum compounds, by Karasek and Karasek[238] in 1911, concerned photographic workers who suffered from severe upper and lower respiratory tract and skin disorders. Similar clinical manifestations—rhinitis, conjunctivitis, asthma, urticaria, and contact dermatitis—have since been reported mainly in chemists and workers engaged in the refining of platinum.[192,218,253,312,322,364] Until recently, such subjects provided the only clinical material for the study of allergy to platinum compounds, and they are still the major source. However, reports by Khan et al.[241] in 1975 and others of anaphylactic reactions in patients under treatment with antitumor platinum coordination compounds[377] now provide an additional source. The findings in these subjects are guides to the investigation of the role of platinum compounds in allergic sensitivity and to the planning of prospective serial studies involving the ordinary population in case environmental exposure becomes significant. It appears that platinum compounds are unique in their ability to cause allergic reactions; compounds of the other platinum-group metals (with the possible exception of rhodium) do not show such activity.

The assumption that the clinical manifestations are due to allergic responses, rather than toxic or irritant effects, is based on three traditional criteria: the appearance of sensitivity is preceded by previous exposure without apparent effect; only a fraction of exposed subjects react to exposure and show evidence in the form of results of skin and provocation tests; and the affected subjects show increasingly high degrees of sensitivity to amounts that are almost always far below those encountered at work or under ordinary circumstances. It is necessary to distinguish between environmental exposure, which is regarded as unlikely to cause sensitization, and the infinitely smaller doses that are capable of eliciting reactions in already-sensitized subjects. Furthermore, prolonged clinical exposure to small doses of allergen may sensitize people, as has been shown in some forms of extrinsic allergic alveolitis, and it is not improbable that this applies as well to asthma.

With respect to many chemical agents, only a low proportion of exposed subjects show evidence of sensitivity. By contrast, the incidence of sensitivity to platinum compounds among refinery workers can be very high, showing the allergenic potency of complex platinum salts, and it can develop rapidly. In one factory,[218] 71% of 91 employees were affected, 57% with asthma and 14% with skin manifestations; in another factory,[312] 65% of the subjects were affected. A 5-year survey[364] of 21 subjects showed all to be affected, with about 40% regarded as asymptomatic, but showing conjunctivitis and nasal mucosal changes, and about 60% having respiratory tract or skin manifestations or both. In most cases, complete cessation of exposure is followed by disappearance of the clinical disorders, presumably because the causal agents are unlikely to be encountered in ordinary life.

There is only scanty immunopathologic, but very suggestive clinical, information on the mechanisms likely to be responsible for the different forms of clinical reaction to the platinum compounds. In the absence of satisfactory *in vitro* immunologic tests, direct tests performed on the sensitized subjects themselves serve in this respect as models. The analysis of the allergic reactions has to be based, by analogy, on the four main types of allergic reaction.[108] These may serve as guides to relevant test procedures already available or urgently requiring development.

TYPE I ALLERGY: IMMEDIATE, ANAPHYLACTIC

The Type I allergic reaction is immediate and is mediated by antibodies that can sensitize mast cells and basophils. This is the most striking and likely to be one of the most informative types of allergic reaction for investigation, not only of occupational, but also of nonoccupational environmental exposure. Clarification of various immunologic terms and assessment of how it should be sought are therefore essential. Their relevance to platinum sensitivity is illustrated below.

Type I, IgE-Mediated Allergy

The production of IgE antibody is characteristic of, although not confined to, a group in the population, termed "atopic," who readily become sensitized by ordinary exposure to common environmental allergens and who are more likely to do so on exposure to occupational agents, also. This group of subjects can serve as biologic monitors of environmental allergens, hence their importance in this and similar contexts.

The term "atopy" is, however, used in a variety of ways: to describe one or more of the clinical manifestations, to indicate the presence of Type I, IgE allergy, or some combination of both. Not only are there differences of opinion on the definition of the clinical entities—for example, asthma[339]—but they may or may not be associated with the presence of IgE antibody.

Use of the term "atopy" to describe the immunologic reactivity of the subject, irrespective of the presence or absence of clinical manifestations, would emphasize the outstanding feature of these subjects, however defined, and would provide a generally acceptable criterion for classification of subjects as atopic or nonatopic.[319] The main proviso here is that the test procedures—skin and serologic—should be such as to give the most specific information.

Skin-Test Procedures

Their high allergenic potency makes it necessary to take great care with platinum salts, as shown by severe reactions to scratch and intracutaneous tests.[148,253,456]

The main methods for skin-testing for Type I allergy are the prick test, the scratch test, and the intracutaneous test. The other methods of testing may suffice for some clinical purposes; but, because it affords the greatest scientific accuracy and for other reasons, the prick test is the most acceptable, being the simplest, the safest, and the most precise. It is performed by pricking with a gentle lifting motion through a drop of test solution into the superficial epidermal layers, with a separate fine hypodermic needle for each preparation. When it is carefully done, and preferably on the flexural surface of the forearm, no blood is drawn and no reactions are elicited at control or negative sites, except in subjects with dermographism. This makes it possible to read even very small wheals as unequivocally positive. These reactions should have, in addition to the wheal, an erythematous flare and are often accompanied by itching. The painless simplicity of the method makes it easy to do a series of tests, starting with the weakest concentrations and repeating doubtful tests rapidly in duplicate or triplicate, as desired, to ensure a confident interpretation. Repetition of tests for this reason is seldom necessary in practice.

Reactions to the prick test show the best correlation with tests for specific IgE antibody in the serum and, when there are symptoms, with the history and provocation tests.[427] Prick tests also reduce considerably the production of nonspecific reactions due to the trauma of testing or to the effects of histamine-liberating agents. A comparison of the threshold concentrations required to elicit wheal reactions to histamine and to histamine-liberators shows that the prick test, which introduces 3×10^{-6} ml into the skin,[424] requires a concentration 1,000–10,000 times greater than that for the intracutaneous test, which introduces an estimated 0.02 ml into the skin. Concentrations of extracts of common allergens are usually 10–100 times higher for prick tests than for intracutaneous tests. This still leaves considerable leeway for avoiding nonspecific wheal effects of test preparations of common allergens or other agents.

The exquisite sensitivity of the test is shown by the positive reactions to prick tests that introduce into the epidermis 3×10^{-6} ml of platinum salt at a concentration of 10^{-9} g/ml.[99,322] This gives an absolute skin-test dose of 3×10^{-15} g of the test compound! This degree of sensitivity emphasizes the need for test procedures that avoid nonspecific reactions, whatever their source, and the care required for excluding even very slight contamination of inactive platinum compounds with the highly allergenic complex platinum salts. The production of urticaria in some sensitive subjects is probably due to the local effects of the platinum salts on the skin and to systemic effects of inhaled or ingested platinum salts.

Scratch tests are not as suitable as prick tests, because they produce more traumatic nonspecific wheals and are more difficult to standardize. With the prick test,[364] it was found that most unexposed persons react to sodium chloroplatinate at 0.01–0.1 g/ml, and many to 0.001 g/ml. This salt has been reported[389] to be the most potent histamine-liberator[312] of the complex platinum salts.

Prick Tests and Atopic Status

Evidence of specific IgE antibody to common relevant environmental allergens is strongly suggested by wheal reactions to carefully performed prick tests; this suggestion is confirmed by such tests as the radioallergosorbent test (RAST) for specific IgE antibody in the serum. In practice, prick tests with a small battery of extracts of common

Allergy to Platinum Compounds

allergens suffice to classify subjects for epidemiologic purposes into atopic and nonatopic immunologic groups, as defined[319] earlier.

A double-blind, statistically controlled study of repeated prick tests with a battery of common allergens at monthly intervals for up to a year in a group of about 50 asthmatic subjects showed no evidence of the induction of sensitivity to extracts that initially gave negative tests (C. W. Clarke and J. Pepys, personal communication). There was a remarkable consistency of positive reactions to the relevant allergens, with, if anything, a tendency for reduction in the sizes of the wheals. These findings are pertinent to the use of such tests for immunologic screening of the population (atopic versus nonatopic) and to the epidemiologic use of repeated serial tests.

The main extracts used so far have been of house dust; of the house-dust allergy mite *Dermatophagoides*, either *D. pteronyssinus* or *D. farinae;* of pollen; and of moulds and animal danders. As might be expected, the subjects who have positive reactions to one or more of these tend to become sensitized to other allergens more rapidly and more frequently. This has been shown in workers engaged in the manufacture of biologic detergents, in whom the chances of Type I sensitization were twice as high in atopic as in nonatopic subjects.[168,300,383]

With respect to platinum-refinery workers, a similar association was reported[364] in subjects with a history or family history of hay fever, asthma, urticaria, and contact dermatitis. In workers of different ethnic populations in two refineries, one in the United Kingdom and the other in South Africa (I. Webster, personal communication), the relationship of atopic status (with regard to prick-test reactions to common allergens) to the appearance of positive reactions to 10^{-9}- to 10^{-3}-g/ml solutions, mainly of $(NH_4)_2[PtCl_6]$, is shown clearly.

In a prospective study of 212 U.K. workers, 64 were regarded as atopic, having had positive prick-test reactions to one or more of the common allergens (the remaining 148 were regarded as nonatopic). None reacted to the platinum salts at 10^{-3} g/ml before starting their employment. Later, a subgroup of 50 workers who had positive prick-test reactions was subjected to further study. Of these 50, 40 also had clinical symptoms of allergy to the platinum salts. Twenty of the 50 were atopic. Thus, 20 of the original 64 atopic subjects had become sensitized, compared with 30 of the 148 nonatopic subjects; this difference is significant at the 1% level. Another nine subjects had negative reactions to prick tests, but had symptoms suggestive of allergy to the salts. The high degree of sensitivity in such subjects, 18 of whom were tested further with other salts, is shown in Table 7-1;[99] many of these and others reacted to concentrations of $10^{-7}-10^{-9}$ g/ml.

TABLE 7-1 Prick-Test Results[a]

Platinum Complex	Test Ser.	A	B	C	D	E	F	G
$(NH_4)_2[PtCl_6]$	(a)	$10^{-4}(3)$	$10^{-5}(3)$	$10^{-6}(3)$	$10^{-7}(+)$	$10^{-5}(2)$	$10^{-5}(2)$	$10^{-5}(2)$
	(b)	10^{-5}	10^{-6}	$10^{-7}(2)$	10^{-8}	10^{-4}	10^{-5}	$10^{-7}(2)$
	(c)			10^{-5}	$10^{-8}(3)$	$10^{-6}(2)$		10^{-5}
	(d)			10^{-7}	$10^{-9}(4)$			
$(NH_4)_2PtCl_4$		10^{-6}	10^{-6}	10^{-5}	10^{-8}	10^{-4}	10^{-5}	10^{-6}
$(NH_4)_2[PtBr_6]$				$10^{-5}(3)$	$10^{-6}(3)$	$10^{-3}(4)$		$10^{-4}(2)$
$Cs_2[Pt(NO_2)Cl_3]$	(a)	$10^{-4}(+)$	$10^{-5}(+)$	$10^{-4}(2)$	$10^{-6}(1)$			
	(b)			$10^{-5}(2)$				$10^{-5}(2)$
	(c)			$10^{-5}(4)$	$10^{-7}(4)$	$10^{-4}(4)$		
$Cs_2Pt[NO_2]_2Cl_2]$		$10^{-3}(\pm)$	$10^{-4}(2)$	$10^{-4}(2)$	$10^{-5}(2)$			
$Cs_2[Pt(NO_2)_3Cl]$	(a)	$10^{-3}(\pm)$	$10^{-3}(2)$	$10^{-3}(1)$	$10^{-4}(3)$			
	(b)			$10^{-4}(3)$	$10^{-6}(2)$	0	0	0
$K_2Pt(NO_2)_4$						0	0	$10^{-4}(2)$
$[Pt(NH_3)_4][Pt(NH_3)Cl_3]$				$10^{-4}(\pm)$				
$cis[Pt(NH_3)_2Cl_2]$								
$trans[Pt(NH_3)_2Cl_2]$								0
cis-$[Pt(CH_3NH_2)_2Cl_2]$				0				0
$trans$-$[Pt(CH_3NH_2)_2Cl_3]$				0				
$[Pt(OHEtNH_2)_2Cl_2]$								
cis				0	$10^{-3}(3)$	0		
trans					0			
cis-$[Pt(NH_3)_2(H_2O)_2](NO_3)_2$					0			
$[Pt(dien)X]X$								
X = Cl					$10^{-3}(\pm)$			
X = Br				0	$10^{-3}(\pm)$	0		
X = I					0			
$[Pt(NH_3)_4]Cl_2$						0	0	0
$[Pt(tu)_4]Cl_2$						0	0	0
$Cs[Pt(gly)Cl_2]$				$10^{-3}(2)$				

Notes: I. (a), (b), (c), (d) for $(NH_4)_2[PtCl_6]$ represent the results of different control tests in the order in which they were carried out. II. For $Cs_2[Pt(NO_2)Cl_3]$ and $Cs_2[Pt(NO_2)_3Cl]$, (a), (b), (c) represent different samples, prepared and purified separately. *On this occasion the subject had taken an antihistamine preparation prior to the test.

tu = thiourea, $(NH_2)_2CS$.

OH Et NH_2 = ethanolamine. OH C_2 H_4 NH_2.

gly = glycine anion, $H_2C(NH_2)COO^-$.

[a]Reprinted with permission from Cleare et al.[99]

Allergy to Platinum Compounds

TABLE 7-1 (Cont.)

H	I	J	K	L	M	N	O	P	Q	R
$10^{-6}(2)$	$10^{-8*}(\pm)$	$10^{-5}(1)$	$10^{-6}(2)$	$10^{-5}(1)$	$10^{-7}(1)$	$10^{-5}(2)$	$10^{-6}(3)$	$10^{-6}(3)$	0	0
$10^{-6}(3)$	$10^{-9}(2)$		$10^{-6}(3)$	10^{-4}	$10^{-7}(3)$				0	0
10^{-7}	10^{-7}		10^{-7}							
	$10^{-9}(8)$		$10^{-6}(4)$							
10^{-7}	10^{-7}		10^{-7}	10^{-4}						
$10^{-4}(\pm)$	$10^{-7}(2)$		$10^{-4}(\pm)$		$10^{-5}(3)$	$10^{-4}(1)$		$10^{-4}(4)$	0	0
	$10^{-4*}(+)$	$10^{-3}(3)$	$10^{-4}(1)$	$10^{-3}(2)$	$10^{-5}(3)$	$10^{-4}(1)$				
$10^{-5}(1)$	$10^{-7}(3)$		$10^{-6}(3)$							0
					$10^{-6}(4)$			$10^{-5}(3)$		
	$10^{-4*}(+)$	$10^{-3}(2)$	$10^{-3}(3)$	$10^{-3}(\pm)$						
	$10^{-3*}(3)$	$10^{-3}(2)$	$10^{-3}(3)$	$10^{-3}(\pm)$						
					$10^{-3}(3)$			$10^{-3}(3)$		0
0										
$10^{-4}(1)$	$10^{-5}(3)$		$10^{-4}(2)$		$10^{-4}(2)$	$10^{-3}(2)$				
	0	0	0	0					0	
	0	0	0	0						
0	0		0		0	0		0	0	
0	0		0		0	0		0	0	
			0		0		0			
	0		0		0		0			
	0		0		0		0			
	0		0		0		0			
	0		0		0		0			
	0		0		0		0			
0										
0										
	$10^{-3}(2)$	0	0	0						

In the other refinery, of 169 subjects, including some who had previously been engaged in the industry, 39 had positive prick-test reactions to $(NH_4)_2[PtCl_6]$ or $Na_2[PtCl_6]$; 17 of these were atopic, and 22 nonatopic. The incidence of atopy in the total group is not known. Of the 17 atopic subjects, six were skin-test-sensitive after 1 month, five after 2–3 months, two after 5–8 months, and four later than this. Thus, in 11 of the 17, sensitization was evident by the fifth month. Of the 22 nonatopic subjects, sensitization was evident in six by the end of 5 months. Only one nonatopic subject showed sensitization within the first 2 months. The remainder required periods of exposure longer than 5 months for sensitization. The difference in the incidence of sensitization between the atopic and nonatopic subjects at 2 months was highly significant; at 5 months, it was significant at the 5% level. Of people with positive prick-test reactions to $(NH_4)_2[PtCl_6]$, one reacted to 10^{-7} g/ml, 10 to 10^{-5} g/ml, seven to 10^{-4} g/ml, and three to 10^{-3} g/ml. Of those who reacted to $Na_2[PtCl_6]$, one reacted to 10^{-7} g/ml, 13 to 10^{-5} g/ml, five to 10^{-4} g/ml, and five to 10^{-3} g/ml.

Analysis of Prick-Test Reactions to Platinum Compounds

The high degree of prick-test sensitivity to the complex platinum salts in sensitized workers made possible an analysis by Cleare et al.[99] of the allergenic relationships of a number of different compounds.

The test materials consisted of a wide range of halide, nitro, and amine complexes (Table 7-1). These were prepared by established methods and had satisfactory elemental analyses. Because refinery workers tested were all very sensitive to $[PtCl_6]^{2-}$ or $[PtCl_4]^{2-}$, the presence of these species as impurities at very low concentrations could give a positive test for an otherwise innocuous material. For example, contamination, not detectable by elemental analysis, of 10^{-3} could elicit reactions at concentrations of 10^{-3} of the test material in subjects capable of reacting to the contaminant itself at 10^{-6} g/ml.

The results show that an allergic response is elicited by charged complexes containing at least one chloro ligand and that, allowing for the inaccuracies inherent in the measurement of wheal reactions, allergenicity is related to the number of chloro groups in the molecules, as shown by the following allergenicity ranking:

$$(NH_4)_2[PtCl_6] \sim (NH_4)_2[PtCl_4] > Cs_2[Pt(NO_2)Cl_3]$$
$$> Cs_2[Pt(NO_2)_3Cl_2] > Cs_2[Pt(NO_2)_3Cl] > K_2[Pt(NO_2)_4];$$

the last compound was inactive.

Amine complexes showed the same general pattern. An important difference is that both *cis* and *trans* dichloro complexes—$[Pt(A)_2Cl_2]$, where A = NH_3, $MeNH_2$—are neutral and give no reactions. The only exception is *cis*-$[Pt(OHEtNH_2)Cl_2]$, which occasionally elicits a reaction at high concentration. This compound is very soluble, in comparison with the other neutral species, and may suffer from slight contamination; 0.001% of $[PtCl_4]^{2-}$ would be sufficient to cause a positive response. A charged species containing two *cis* chloro ligands, $[Pt(gly)Cl_2]$, is slightly active. Amine species with one chloro group are either slightly active or inactive, irrespective of charge, although contamination may be a problem. The fully substituted amine $[Pt(NH_3)_4]Cl_2$ is totally inactive, as is $[Pt(tu)_4]Cl_2$, another complex in which all the coordinated chlorides have been replaced.

Changing from chloro to bromo ligands yields complexes that are less physiologically active, so that the order $(NH_4)_2[PtCl_6] > (NH_4)_2[PtBr_6]$ is observed in all cases. $[Pt(dien)Br]Br$ and $[Pt(dien)I]I$ are inactive, as is the corresponding chloro complex.

Allergy to Platinum Compounds

The reactivity of the complexes depends largely on the leaving ability of the ligand that is being replaced. Table 7-2 shows the general conclusions that can be drawn from these results. Reactions 1 and 2 (Table 7-2) are probably most relevant, in that coordination positions other than the leaving group under study are occupied by an inert trichelating ligand. Thus, such leaving groups as the halogens are consistently fairly labile and reactive, whereas nitro, $-NO_2^-$, and thiocyanato, $-SCN^-$, groups are more inert and much less reactive. The platinum–amine linkage is very stable and inert to nucleophilic attack; i.e., amines are very poor leaving groups.

The results of the allergy tests can be explained on this kinetic basis. Complexes containing strongly bound ligands with poor leaving abilities are not allergenic, presumably because there is little or no reaction with proteins—e.g., $[Pt(NH_3)_4]Cl_2$, $K_2[Pt(NO_2)_4]$, and $[Pt(tu)_4]Cl_2$. Allergenic activity increases with the number of halide ligands, the chloro complexes being more effective than bromide, as forecast by Reaction 2.

TABLE 7-2 Relative Lability, $k_2(X)/k_2(N_3)$, of Ligands Displaced from Platinum(II) Complexes[a]

$[Pt(dien)X]^+ + py \rightarrow [Pt(dien)(py)]^{2+} + X^-$ (water, 25° C)							(1)
Cl >	Br >	I >	N_3 >	SCN >	NO_2 >	CN	
40	27	12	1.0	0.36	0.056	0.02	
$[Pt(dien)X]^+ + Y^- \rightarrow [Pt(dien)Y]^+ + X^-$ (water, 30° C)							(2)
I ~	Br >		Cl >		N_3	(> NO_2)	
73	70		33		1.0	?	
$[Pt(bipy)(NO_2)X] + Y^- \rightarrow [Pt(bipy)(NO_2)Y] + X^-$ (methanol, 25° C)							(3)
I >	Br >		Cl >		NO_2 >	N_3	
900	240		140		7	1.0	

[a] Data from Cleare et al.,[99] Basolo et al.,[33] Belluco et al.,[42] and Cattalini and Martelli.[81]

dien = diethylenetriamine, $H_2N \cdot CH_2 \cdot CH_2 \cdot NH \cdot CH_2 \cdot CH_2 \cdot NH_2$.
py = pyridine, C_5H_5N.
bipy = 2,2'-bipyridyl,

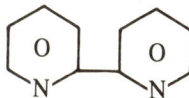

Three postulates are examined for the platinum–protein interaction. Protein reaction[424,476] with more than one platinum coordination position, such as protein bridging, is unlikely, because compounds with only one reactive group are allergenic.[99] The antibodies elicited by exposure to $[PtCl_4]^{2-}$ seem capable of reacting with different structures within the approximately planar coordination around the platinum(II) center attached to a protein or other macromolecular carrier.

The fact that platinum(IV) and platinum(II) chloro species are equally allergenic in the same individual may be due to reduction of platinum(IV) to platinum(II), as postulated for the anticancer agents. Platinum(IV) species are usually kinetically inert, unless there is a reaction through a platinum(II) intermediate. The possibility that platinum(IV) will react in its own right in such a highly sensitive biologic system is slight, because platinum(II) is not much more allergenic than platinum(IV). However, one cannot rule out the possibility that the sensitivity to platinum(IV) compounds observed in the refinery workers was due to their working under refinery conditions where they were initially sensitized to platinum(IV).

Prick tests were made on one highly sensitive subject who showed positive reactions to $(NH_4)_2[PtCl_4]$ and $(NH_4)_2[PtCl_6]$ at 10^{-9} g/ml. Additional tests were made on the same subject with equimolar conjugates of these salts containing human serum albumin, transferrin, and ovalbumin and with a mixture containing the salt and a conjugate of human gamma globulin in a 3:1 molar ratio of protein to platinum salt. The platinum content of these conjugates ranged from 6 to 27 ppm (by weight). All produced *negative* reactions; this raised questions about the nature of the carrier protein when the salts themselves are used for testing and about whether the spatial arrangements of the platinum ions are such as to enable bridging of IgE antibody molecules on the surface of the skin mast cells. Ishizaka and Ishizaka[228] found that 7 S molecules with a molecular weight of about 30,000 are necessary for this bridging, which triggers the release of histamine from the cells. It is conceivable that the platinum conjugates do not fulfill this requirement.

In summary, the allergenicity of the platinum compounds in the refinery subjects is confined to a small group of charged compounds containing reactive ligand systems (such as chloride and, to a lesser extent, bromide) and is directly related to both their charge and their overall reactivity. The refinery workers sensitized to specific platinum complexes apparently do not react to the majority of platinum compounds.

Allergic Reactions to a Platinum-Coordination-Compound Antitumor Agent

The report by Khan et al.[241] of an anaphylactic reaction 9 months after the seventh injection of cis-dichlorodiammineplatinum(II) (DDP) raises other possibilities. The patient had immediate (Type I) reactions to DDP, to cis-diiododiammineplatinum(II), $PtI_2(NH_3)_2$, and to potassium chloroplatinate, $K_2[PtCl_6]$, and potassium chloroplatinite, $K_2[PtCl_4]$. Positive reactions were also elicited with the diaquo derivative of DDP, $Pt(NH_3)_2(H_2O)_2^{2+}$; this indicates that halogen ligands are not necessary. In the tests by Cleare et al.,[99] DDP and its diaquo derivative did not produce reactions in refinery workers who were highly sensitive to $(NH_4)_2[PtCl_6]$ and $(NH_4)_2[PtCl_4]$. The patient reported by Khan et al.[241] did not, however, show sensitivity to some square-planar complexes, such as platinum blue, platinum(II)-1,2-diaminocyclohexane malonate, and platinum(II)-ethylenediamine malonate. Significant histamine release was obtained on incubation of the patient's leukocytes with DDP. A raised serum content of IgE was found, as has also been the case in three of 22 sensitized refinery workers.

The differences between the refinery workers and the affected patient may be attributable to differences in their modes of sensitization and the chemical agent responsible. The possibility of contamination with minute amounts of a complex platinum salt of the therapeutic platinum coordination compounds has to be considered, but this would not explain the absence of reaction to DDP, etc., in the refinery workers, nor would it account for the positive reactions to DDP, diaquo-DDP preparations, and other coordination compounds, as well as to the complex platinum salts, in the affected patient.

Radioallergosorbent Tests for Specific IgE Antibodies to Platinum Compounds and Conjugates in Sensitized Refinery Workers

Radioallergosorbent tests[33,42,81,228,230,476,477] (RAST) have been used with a wide range of conjugates for the presence of an IgE antibody specific to platinum conjugates. In spite of the high degree of Type I skin-test reactivity to the platinum salts, it has not been possible to demonstrate IgE antibodies.

The agents used so far for linking to paper disks for the RAST include conjugates of $(NH_4)_2[PtCl_4]$ and $(NH_4)_2[PtCl_6]$ with human, bovine, and rabbit serum albumin, ovalbumin, gelatin, and collagen; bovine gamma globulin; whole dialyzed human and rabbit serum; albumin-depleted human serum; and human skin sections. Molar ratios of coupling of platinum to protein have ranged from 10:1 to 1:100.

Platinum compounds were also linked to Sepharose (Pharmacia) particles via an amino acid derivative. These included $K_2[PtBr_2(NO_2)_2]$, $Cs_2[Pt(NO_2)_2Cl_2]$, $(NH_4)_2[PtBr_6]$, cis-$Pt(NH_2CH_3)_2Cl_2$, $(NH_4)_2[PtCl_4]$, and $(NH_4)_2[PtCl_6]$.

Passive-Transfer Tests in Man and Monkey

The report[13] of positive reactions to passive-transfer tests in a nonsensitive human recipient, in whom positive reactions were elicited by tests of the sensitized sites with platinum conjugates, suggests the presence of specific IgE antibody.

Passive-transfer tests (with unheated and heated serum of 10 sensitive subjects) for the presence of specific IgE and short-term specific IgG antibodies were made in three monkeys that were given intravenous injections of $Na_2[PtCl_6]$, $Na_2[PtCl_4]$, and $(NH_4)_2[PtCl_6]$ with Evans blue (W. E. Parish and J. Pepys, personal communication; Parish[310,311]). All the tests were negative.

The difficulty of demonstrating IgE antibodies specific for the small-molecule substances, such as platinum, even when conjugated to carrier proteins is common in allergy to small-molecule drugs. In the case of penicillin allergy, conjugates of penicilloyl determinants with polylysine and serum proteins have been effective in RAST for demonstrating specific IgE antibody,[477] but this is an exception, rather than the rule. Because specific IgE antibody, if present, against platinum conjugates could constitute one of the most sensitive laboratory tests for sensitization (particularly in atopic subjects) and thus provide evidence of atmospheric contamination with appropriately reactive platinum derivatives, it is urgent that such a test be developed.

Immediate Reactions to Histamine Liberation by Complex Platinum Salts

Immediate reactions to histamine liberation by complex platinum salts are relevant to skin and other tests, but are not likely to be playing more than a subsidiary role in sensitized subjects who can react to very small doses that are not histamine-liberating. Such an effect could be expected in all subjects, irrespective of sensitization.

It has been reported that sodium chloroplatinate, in particular, liberates histamine on intravenous injection into guinea pigs and on addition to guinea pig ileum.[313,389,427] During investigations[192] in which 0.1-ml volumes of 10^{-4} concentrations of complex platinum salts were injected intracutaneously into guinea pigs previously treated with intravenous injections of Evans blue, it was found that the sodium chloroplatinate and chloroplatinite salts caused the most exudation of dye; ammonium chloroplatinate and chloroplatinite caused trace to moderate exudation. Conjugates with human serum albumin of ammonium chloroplatinite, sodium chloroplatinite, and chloroplatinic acid produced only trace exudation of dye. This demonstration that salts other than sodium chloroplatinate may cause histamine liberation on introduction into the skin of guinea pigs makes appropriate controls for such tests necessary.

TYPE II ALLERGY: AUTOALLERGIC

In Type II allergy, the haptene combines with a cell surface to produce the complete antigen against which an immunologic response may be mounted. This is a form of autoallergic reaction. The capacity of the platinum coordination compounds to combine with DNA on the surface of malignant cells and T-lymphocytes[373] makes it necessary to keep this form of allergy in mind as a possibility with platinum compounds.

TYPE III ALLERGY: IMMUNE-COMPLEX, COMPLEMENT-DEPENDENT

The production of precipitating antibodies and their combination with a moderate excess of antigen lead to the formation of toxic soluble complexes that fix and activate the C3 component of the complement. Tissue-damaging Type III reactions ensue from the effects of these enzymatic aggregates.

The possibility of Type III allergic reactions to platinum salts is suggested by the findings of positive passive-transfer reactions in the guinea pig, in which the heterologous human precipitating antibody sensitizes mast cells, so that histamine is liberated on allergen challenge (W. E. Parish and J. Pepys, personal communication).

In these tests, each guinea pig was passively sensitized by intracutaneous injections of 0.05–0.1 ml of the serum of the particular group of subjects and 4 h later was given intravenous injections of Evans blue and 1–1.5 mg of the particular platinum salt or conjugate. In the first series of 10 subjects, the tests were made with $Na_2[PtCl_6]$, with $Na_2[PtCl_4]$, and with $(NH_4)_2[PtCl_6]$. The serum of two of the subjects gave reactions regarded as positive; in one of these, the two specimens were taken at different times. Reactions were elicited with serum of both subjects to $(NH_4)_2[PtCl_4]$ and with serum of one of them to $Na_2[PtCl_4]$ as well. In the second series of nine different subjects, the platinum salts or conjugates were injected intracutaneously into sites passively sensitized 4 h earlier; the guinea pigs had been treated previously with intravenous injection of Evans blue. In this test, serum of three had weak or faint, equivocal reactions to 1-mg test doses of $(NH_4)_2[PtCl_4]$ and $Na_2[PtCl_6]$; the serum of one of them also had a faint reaction to $(NH_4)_2[PtCl_6]$. Tests with conjugates of $Na_2[PtCl_4]$ with human serum albumin and of $(NH_4)_2[PtCl_4]$ with horse serum produced no reactions at all. The fact that serum of three of the nine had these weak and uncertain reactions, whereas all the others were negative, suggests that they may have been positive; but the reactions were not unequivocal, as was the case with the serum from the two guinea pigs. In a third series of tests with serum of 25 guinea pigs, the guinea pig skin was sensitized with 0.1-ml volume of the serum undiluted and at a 1/10 dilution, and the tests were made by intravenous injections of 1 mg of $Cs_2[Pt(NO_2)_3Cl]$, $Cs_2[Pt(NO_2)Cl_3]$, and their conjugates with human serum albumin; cis-$Pt(A_2)Cl_2$ (where $A = CH_3NH_2$); $K_2[Pt(NO_2)_4]$; $Cs_2[Pt(NO_2)_2Cl_2]$; and $(NH_4)_2[PtCl_4]$. No positive reactions were elicited.

These tests, albeit with only limited results, suggest that precipitating antibodies are demonstrable in a small number of the exposed subjects. The production of precipitins requires, as a rule, more intensive antigenic stimulation and continued exposure for their demonstrable persistence in the serum. Many of the test specimens of serum were taken at least months after cessation of exposure, which in any case would probably have been relatively limited.

Further suggestions of the possible presence of precipitating antibodies were provided by a report[253] on hyposensitization of a chemist who was highly sensitive to $(NH_4)_2[PtCl_6]$. A prick test on this patient with an 8% solution of $(NH_4)_2[PtCl_6]$ caused a severe anaphylactic reaction. An attempt was made to hyposensitize him with intracutaneous injections of gradually increasing amounts. When he reached the 5-μg dose, a reaction resembling serum sickness appeared, with urticarial skin eruptions, swelling of joints, and red eruptive papules that showed, histologically, a vasculitis with eosinophil infiltration, as did some of the injection sites at which necrosis also appeared. This reaction is very suggestive of Type III allergy, an important part of serum-sickness reactions. The patient's sensitivity to occupational exposure decreased. Auto-passive-transfer tests (in which serum taken before, during, and after injection treatment was mixed with the platinum salt and then, after incubation, injected intracutaneously into the patient's back) showed decreased reaction to the test with the posttreatment serum mixture, suggesting that blocking antibodies had been formed. Repetition of the hyposensitization treatment at a later date resulted in the same pattern of reactions resembling serum sickness. The patient's

Allergy to Platinum Compounds

clinical sensitivity to $(NH_4)_2[PtCl_6]$ was also decreased, although exposure to $(NH_4)_2[PtCl_4]$ caused eye, nasal, and skin reactions again, which suggested allergenic specificity of the different salts.

Tests for the presence of antibodies showed no response to challenge of his lymphocytes *in vitro* with $(NH_4)_2[PtCl_6]$ or to a conjugate of $(NH_4)_2[PtCl_4]$ in solution. Serologic tests produced negative results for precipitins and hemagglutinating antibodies. Uniformly negative results have also been found in precipitin and hemagglutination tests and in tests with red cells and latex particles linked to a wide range of platinum salts and their conjugates with protein.

Recent studies[129] have shown that a number of extracts of organic substances implicated in Type III allergic lung reactions in man and animals[320] can also activate the C3 component of the complement directly (an alternate pathway). Such possibilities have now to be considered with all agents that give rise to reactions that have features of Type III allergy, including platinum compounds.

With well-established allergens, it can often be claimed confidently on the basis of positive skin-test reactions that the relevant allergen is of clinical importance in the particular subject. Many workers, however, consider that provocation tests are necessary. In spite of the reported hazards of skin-testing with platinum salts,[148,456] it is possible to have not only safe skin tests by the prick-test method, but also nasal tests with solutions of the platinum salts and occupational-exposure tests for asthmatic reactions to the dust that rises when mixtures of small amounts with lactose powder are poured from one receptacle to another.[322]

Nasal Tests

Nasal tests were made[322] with 0.01–0.02 ml of the test material introduced (after a control test with diluent) into one nostril at the lowest concentration and then, if the results were negative, at increasing concentrations. Positive reactions consist of itching of the nose, sneezing, rhinorrhea, and sometimes nasal obstruction. These reactions appear within a minute or two and run the same time course as the wheal reaction to the prick test. Seven of 11 subjects had positive reactions to concentrations of 10^{-5}–10^{-3} g/ml.

Bronchial Tests

The complex salts of platinum were prepared in powder form as separate mixtures containing 40 mg of each salt and 1 kg of lactose powder. The control lactose powder and the mixtures were dried for 20 h at

105° C and then kept in a desiccator, because the lactose is hygroscopic and loses its powdery, dusty consistency.

The test consisted of having the patient shake 250 g of the mixture repeatedly from one tray to another that was 0.3 m below it[322] and inhale the dust so created. Initially, daily tests were made with increasing amounts, in microgram quantities, of the platinum salts mixed with lactose, until it was found that 40 mg in 1 kg of lactose sufficed to elicit reactions in sensitive subjects.

The maximal total exposure was 30 min, divided into segments of 5, 10, and 15 min with 10-min intervals between them. Exposure was terminated at any time if there was evidence of a clinical reaction or decrease in ventilatory function. Tests of ventilatory function were made before the test and at each interval during the exposure, then every 10 min for 1 h after its completion, and then every hour for 8 h or until any reaction had resolved. No further exposure was made if the 1-s forced expiratory volume (FEV_1) decreased by at least 10%, and only one test was made per day. The patients were exposed to control lactose alone and to a mixture of platinum salt and lactose.

The 16 subjects investigated had worked in a platinum refinery—nine for less than 6 months, five for less than a year, and two for many years—before showing signs of sensitivity. The majority were either quite well or much improved on leaving the refinery. Detailed measurements of pulmonary function on admission for investigation showed 10 to be normal in all respects, five to have slight airway obstruction, and one (who was not found to be sensitive to platinum and who was a heavy smoker) to have evidence of early restrictive ventilatory defect with a decrease in carbon monoxide gas-transfer factor. All were investigated many months after (some more than a year after) exposure ceased.

Inhalation tests elicited positive immediate reactions, starting in 10 min and resolving in 1 h, in eight patients, all of whom had immediate prick-test reactions. These reactions were blocked by pretest inhalation of cromolyn sodium, just as immediate asthmatic reactions to test with common allergens are blocked; this suggests a similar mechanism. In one of the eight there was also a late reaction. In two other subjects, only late reactions were elicited: in one, the reaction started after 30 min, reached a maximum at 1.5 h, and resolved in 4 h; in the other, the reaction started after 4 h and reached a maximum at 7 h, when it was reversed by an injection of epinephrine. Neither of these patients had immediate prick-test reactions; in one, the nasal test was also negative. Thus, 10 of the 16 subjects had positive bronchial reactions. Rhinitis was also provoked in six cases, conjunctivitis in three, dermatitis in two, and urticaria in one case; some of these occurred together.

Allergy to Platinum Compounds

The immediate asthmatic reactions are analogous to the Type I skin-test reactions in speed of appearance and in duration, and they are like those produced by common allergens when IgE is involved. The late reactions are like those to common allergens when Type III allergy is thought to be present, and they are compatible with Type III reactions in speed of appearance and in duration.

It must be pointed out that, although inhalation tests conducted in this manner are reasonably reproducible and can establish the *relative* allergenicity of various compounds, they cannot be quantitatively related to airborne concentrations in micrograms per cubic meter. For this reason, they are not very useful in helping to set standards. Furthermore, these tests are not very sensitive measures of the effect of particle size of the allergens, a characteristic that is probably important in determining the degree of respirability of the compounds.

Comparison of Different Platinum Salts

Table 7–3 shows that the salts differ in their capacity to elicit reactions, with ammonium tetrachloroplatinate being the most potent and sodium hexachloroplatinate the least potent. These results show that, with care, it is possible to make skin and provocation tests safely for establishing the presence of allergic sensitization to the platinum salts. They can also be used effectively for identifying the particular allergen or allergens responsible for sensitizing individual patients.

TYPE IV ALLERGY: DELAYED TUBERCULIN-TYPE ALLERGY

Contact dermatitis is part of the skin manifestations of sensitivity to platinum compounds. Type IV allergy has been proposed as the mechanism to explain such reactions, and both skin (patch) and *in vitro* lymphocyte tests can be used for diagnosis of this form of platinum sensitivity. Care is required with skin tests to exclude the presence of Type I allergy, so as to avoid undesirable reactions, and with lymphocyte tests to exclude nonspecific effects of platinum compounds on lymphocytes.

CONCLUSIONS

Investigations for allergic sensitivity to platinum compounds need to be based, until laboratory tests for antibodies and cellular sensitivity are developed, mainly on the use of skin tests for Type I (immediate) allergy. The exquisitely high degree of this type of allergy in which

TABLE 7-3 Comparison of Inhalation Tests with Complex Platinum Salts[a]

Patient	Platinum Salt					
	$(NH_4)_2[PtCl_6]$		$(NH_4)_2[PtCl_4]$		$Na_2[PtCl_6]$	
	Exposure, min	Decrease in FEV_1, %	Exposure, min	Decrease in FEV_1, %	Exposure, min	Decrease in FEV_1, %
1	15	38	10	49	10	35
2	4	28	5	45	30	16
3	6	31	10	38	15	23
4	10	32	10	34	30	0
5	30	0	30	18	30	0
6	30	0	30	15.5	30	0
7	30	26	15	30	30	0
8	15	29	20	29	20	19
9	30	17	30	17	30	0

[a]Data from Pepys et al.[322]

extremely low doses elicit reactions in sensitized platinum-refinery workers and chemists endows it with considerable value as a means of assessing the presence of similar complex platinum salts in the environment and makes it possible to study cross-reactivity between such substances. Patch tests for Type IV (contact dermatitis) allergy, which is likely to be far less common, may also find a more limited role. These investigations are appropriate and are already being applied to refinery workers in whom preemployment and serial tests are valuable in detecting the development of sensitivity at an early stage.

The questions posed here are how these findings can be applied to the investigation of environmental sensitization to platinum compounds in the general population and who may be encountering them in one form or another as new atmospheric contaminants. The evidence of the greatly increased capacity for sensitization (not only to common allergens, but also to platinum salts) of the particular immunologic group in the population termed "atopic" means that people in that group can be regarded as "biologic monitors" of environmental allergenic substances.

In the ordinary clinical investigation of atopic subjects, it is common practice to include routinely, with the relevant common allergens or as additional preparations, tests with other potential allergens. For example, investigators have for the last 5 years used extracts of the enzymes of *Bacillus subtilis* (used in biologic detergents) for routine testing in patients. These materials caused Type I allergy in heavily exposed workers when they were introduced and even in small numbers of so-called consumers.[145,321,323] It was therefore desirable to determine, in communities where these were being used, the presence or absence of allergy to them. Tests of nearly 2,500 patients,[323] now extended to a total of approximately 5,000 patients (Pepys, personal communication), about two-thirds of whom were atopic, showed no evidence of sensitivity, thus demonstrating that they are acceptable in this respect in the form in which they are being used. A similar approach is appropriate with other new potential environmental allergens to provide information on whether Type I allergy (at least to agents tested) is present or is developing in the community with the passage of time. Such tests can be done routinely as patients present themselves for investigation. Selection of test products can be guided by information from sensitized workers and perhaps patients under treatment with platinum coordination compounds. The skin-prick test, if carefully used, is simple, safe, reproducible, and highly precise; it also minimizes problems of nonspecificity due to test methods or irritancy of test materials. The introduction of such tests into selected centers as part of routine investigations could be used as a guide to whether sensitization is occurring and, if so, whether it can be correlated with measurements of platinum compounds in the environment.

Considering the simplicity, reliability, and safety of these skin-prick tests, coupled with the possibility of human exposure to low concentrations of platinum-group metals that may enter the environment through new uses, it seems that failure to include such tests as a routine part of examinations in appropriate clinics would mean overlooking the most sensitive method now available to monitor any changes in human sensitivity to these new environmental pollutants.

8

Environmental Considerations

The purpose of this chapter is to assess the potential impact of *new* sources of platinum-group metals that may provide pathways for their entry into the environment. Before such an assessment can be properly made, it is important that the current distribution of the metals be ascertained. The first section discusses what little information is available on the concentrations in various segments of the environment and in tissue. The major potential contributors of platinum-group metals to the environment as a result of actions of man can be divided into mobile sources and stationary sources. A discussion of these two categories makes up the remaining sections of this chapter.

PRESENT DISTRIBUTION

Soil, Water, and Air

As mentioned in Chapter 2, the concentrations of the various platinum-group metals in the earth's crust are estimated to range from 10 ppb for palladium and 5 ppb for platinum to less than 1 ppb for osmium, iridium, and ruthenium.[269] Such concentrations are far too low to allow economic extraction. Ores in South Africa, Canada, and the U.S.S.R. contain much higher concentrations—1–10 ppm—and these provide the bulk of the world's supply of these materials. Before

TABLE 8–1 Baseline Concentrations of Platinum and Palladium in Surface Soil, Ambient Air, and Water[a]

Type of Sample and Type of Area	Location	Concentration	
		Platinum	Palladium
Soil:			
Near freeway	Lancaster, Ca.	<0.8 ppb	<0.7 ppb
Near freeway	Los Angeles, Ca.	<0.8 ppb	<0.7 ppb
Mining	Sudbury, Ont., near mine	<0.8 ppb	<0.7 ppb
Mining	Sudbury, Ont., precious-metals section	<0.8 ppb	4.5 ppb
Mining	Sudbury, Ont., Copper Cliff mine	<0.8 ppb	2.0 ppb
Air:[b]			
Near freeway	Lancaster, Ca.	$<5 \times 10^{-8}\ \mu g/m^3$	$<6 \times 10^{-8}\ \mu g/m^3$
Near freeway	Los Angeles, Ca.	$<5 \times 10^{-8}\ \mu g/m^3$	$<6 \times 10^{-8}\ \mu g/m^3$
Mining	Sudbury, Ont., engineering building	$<3 \times 10^{-3}\ \mu g/m^3$	$<3 \times 10^{-3}\ \mu g/m^3$
Mining	Sudbury, Ont., south mines	$<3 \times 10^{-3}\ \mu g/m^3$	$<3 \times 10^{-3}\ \mu g/m^3$
Refinery[c]	Precious-metals section	$0.377\ \mu g/m^3$	$0.291\ \mu g/m^3$
Refinery[c]	Furnace room	$<3 \times 10^{-3}\ \mu g/m^3$	$<3 \times 10^{-3}\ \mu g/m^3$
Refinery[c]	Salts section	$0.18\ \mu g/m^3$	$0.03\ \mu g/m^3$
Refinery[c]	Refinery section	$0.16\ \mu g/m^3$	$0.09\ \mu g/m^3$
Water:			
Near freeway	Lancaster, Ca.	<0.08 ppb	<0.024 ppb
Near freeway	Los Angeles, Ca.	<0.08 ppb	<0.024 ppb
Mining	Sudbury, Ont.	<0.05 ppb	<0.015 ppb

[a]Data from Johnson et al.[233] Measurements based on atomic-absorption spectrophotometry.
[b]OSHA standard, $2.0\ \mu g/m^3$ (as soluble salts).
[c]"Refinery "B," Johnson–Matthey, New Jersey.

Environmental Considerations

discovery of the palladium–platinum ore in Montana, there were no areas in the United States (except for some deposits near Goodnews Bay, Alaska[234]) where the concentration was thought to be sufficiently high to allow economic primary mining of the platinum-group metals.

Because of the very low concentrations, only a few measurements of these metals have been reported. The most reliable data appear to be those collected by the Southwest Research Institute (SWRI) on platinum and palladium in 1974 (before the introduction of cars with catalytic converters) to establish a baseline for comparison with future measurements.[233] The data they collected on surface-soil, ambient-air, and water samples are given in Table 8-1. The measurements were made with a carefully calibrated Perkin–Elmer 306 atomic-absorption spectrophotometer. The surface-soil samples analyzed (except for two samples taken in the precious-metals section of the Canadian mines) were below the detection limits of 0.8 and 0.7 ppb for platinum and palladium, respectively. The air samples (except those in the refinery and salts sections of a typical refinery in New Jersey) were also below the detection limits of 5×10^{-8} and 6×10^{-8} $\mu g/m^3$ for platinum and palladium, respectively (California samples), and 3×10^{-3} $\mu g/m^3$ for either platinum or palladium (mine, refinery data). It might be noted that the air samples taken in the Canadian mines were in the presence of operating fork-lift trucks equipped with catalytic converters that contained noble metals and were similar to those now on many U.S. automobiles. Apparently, there was no detectable emission of platinum or palladium from these devices. All these measurements are well within the OSHA standard of $2 \mu g/m^3$ (as soluble salts) described in the *Federal Register*.[452] Finally, the same group analyzed tapwater in southern California and water in and around the Sudbury, Ontario, Canadian mines. Again, they found no measurable amounts of the platinum-group metals in any of the samples.

On the basis of these very sparse data, it is impossible to make a sweeping generalization about the concentrations of platinum-group metals in soil, air, or water. However, it is highly unlikely that harmful concentrations of these species exist in air, water, or soil anywhere in the United States (except possibly within the confines of the two noble-metal refineries on the East Coast). Additional data are needed to test this prediction.

Data do not appear to be available with respect to coal. As more and more coal is consumed in the United States, it becomes important to assess any potential hazards that might be associated with trace components, including the platinum-group metals. It is recommended that data be collected to measure the concentration of these materials in coal and to determine their fate as coal is consumed in combustion, synthetic-fuel production, and chemical synthesis.

Only recently have baseline data begun to appear on the four minor platinum-group metals (rhodium, ruthenium, osmium, and iridium). Their concentrations are below the detection limits of all but the most sensitive instruments. Data on ruthenium, which is a byproduct of nuclear waste, have been reported by Brown.[63] Although no dramatic increase in use of these metals is anticipated (except possibly rhodium and ruthenium in catalytic converters), it is important that this type of information be compiled, if for no other reason than to allow one to assess their role in biologic processes.

Additional data can be obtained from the book by Bowen[59] and the report by Dawson.[118]

Man

With the increased possibility of human exposure to platinum-group metals through escape of auto emission control catalysts or through use of antitumor chemotherapy agents containing the metals, there has been some activity recently to establish baseline data for the body burden of these materials. The potential hazard is outlined in a review by EPA researchers.[66]

As noted in Chapter 6, studies with radioactively labeled platinum, palladium, iridium, and ruthenium indicated that the quantities of these metals not secreted in the urine or feces are retained mainly in the kidneys, spleen, and liver.[397] It might therefore be expected that their concentrations are greater in these organs than in other organs.

Two careful studies have now been completed to measure the concentrations of platinum and palladium in various tissues, but the data appear to be quite contradictory. The SWRI group responsible for the data in Table 8-1 collected autopsy tissue samples from 10 people 12-79 years old who died from a variety of causes in southern California; the data are given in Table 8-2.[233] For all samples analyzed, the concentrations were below the limit of detection by atomic-absorption spectrophotometry—i.e., less than about 1-10 ppb. Similar observations were made on blood, urine, hair, and feces collected from 282 people living in southern California, as indicated by the data in Table 8-3. A composite of all the blood samples indicated concentrations of approximately 0.1 and less than 0.01 μg/100 ml for platinum and palladium, respectively. However, because a number of uncertainties are associated with measurements of such large samples (about 750 ml), these results are probably reliable only to within a factor of 2 or 3. Nevertheless, they suffice to indicate the extremely low concentrations

Environmental Considerations

TABLE 8-2 Platinum and Palladium in Autopsy Samples, Los Angeles[a]

Tissue	Concentration, ppb (wet tissue)	
	Platinum	Palladium
Liver	<0.24	<0.6
Kidney	<2.6	<6.7
Spleen	<1.3	<3.3
Lung	<1.3	<3.3
Muscle	<0.9	<2.2
Fat	<1.3	<1.6

[a]Data from Johnson et al.[233] Measurements based on atomic-absorption spectrophotometry. Number of autopsy cases, 10. Ages, 12-79 years. Sex, five male and five female. Causes of death, pancreatic carcinoma, laryngeal carcinoma, aplastic anemia, urinary bladder carcinoma, acute lymphocytic leukemia, cervical adenocarcinoma, hypertension/hypoproteinemia sepsis, myocardial infarction, and hypertensive cerebrovascular accident.

of platinum and palladium in the blood of people living in southern California.

The second set of measurements, for platinum only, was made by Stewart Laboratories[125] on autopsy tissue samples from 97 people, 95 from southern California and two from New York City. Up to 21 tissue samples were collected from each cadaver for analysis (1,303 samples in all). Sixty-two samples from 45 cadavers had detectable concentrations (>10 ppb for a 1-g wet sample) of platinum, the detectable concentration being $0.003-1.43\,\mu g/g$ of wet tissue (mean, $0.16\,\mu g/g$). Table 8-4 shows the frequency of detection of platinum in the various tissue samples. The authors contended that the frequency of occurrence is a measure of the distribution of platinum among the various body organs.

Why the Stewart Laboratories values are considerably higher than those obtained by the SWRI researchers is not obvious. However, as a general rule, when measuring amounts of trace substances where unintentional contamination is possible, it is usually safer to accept the *lower* value as the more accurate. The Subcommittee on Platinum-Group Metals tends to apply this criterion in the present case and thus put more reliance on the SWRI data. Furthermore, the statistical occurrence of platinum in the various tissues in Table 8-4 is probably not a significant indication of its distribution in humans.

Autopsy data have been collected on two cancer patients who died at the Wadley Institute of Molecular Medicine. The earlier set has been published,[199] but the more complete set (R. J. Speer and H. Ridgway,

TABLE 8–3 Platinum and Palladium Concentrations in Southern California[a]

Group	Concentration in Blood, μg/100 ml		Concentration in Urine, μg/liter		Concentration in Hair, μg/g		Concentration in Feces, μg/g	
	Platinum	Palladium	Platinum	Palladium	Platinum	Palladium	Platinum	Palladium
Age:								
1–16 years	<3.1	<0.9	<0.6	<0.3	<0.05	<0.02	<0.002	<0.001
17–34 years	<3.1	<0.9	<0.6	<0.3	<0.05	<0.02	<0.002	<0.001
35+ years	<3.1	<0.9	<0.6	<0.3	<0.05	<0.02	<0.002	<0.001
Composite:								
Los Angeles	0.049	<0.01						
Lancaster	0.18	<0.01						

[a] Data from Johnson et al.[233] Measurements based on atomic-absorption spectrophotometry.

TABLE 8-4 Occurrence Frequency of Tissue Samples with Detectable Platinum[a]

Tissue	No. Tissue Samples Analyzed	Samples with Detectable Platinum	
		No.	%
Subcutaneous fat	74	10	14
Kidney	91	11	12
Pancreas	84	10	12
Liver	90	10	11
Brain	9	1	11
Gonad	53	5	9
Adrenal	60	3	5
Muscle (psoas)	97	4	4
Aorta (descending)	92	3	3
Heart (left ventricle)	82	2	2
Spleen	52	1	2
Prostate/uterus	63	1	2
Thyroid	73	0	0
Lung	95	0	0
Vertebra (lumbar)	94	0	0
Rib (fifth)	97	0	0
Femur	57	0	0
Clavicle	30	0	0
Hair, scalp	9	0	0
Hair, pubic	1	0	0
	1,303	61	—

[a] Data from Duffield et al.[125] Measurements based on emission spectrochemistry. Number of autopsy cases, 97—95 from California and two from New York City. Average age, 61.5 years. Sex, 39 male, 52 female, six unknown.

personal communication) is still unpublished; both are contained in Table 8-5. Some of these values are at least an order of magnitude higher than the highest values reported by Stewart Laboratories and may reflect contamination or treatment of the patient with chemotherapeutic agents containing platinum complexes (although the researchers claimed that neither patient received these drugs). Part of the discrepancy lies in the concentrations' being reported in terms of *wet* tissue by the Stewart and SWRI researchers, whereas the Wadley data refer to *dry* tissue. The percentage of water varies from one tissue to another, and the value is somewhat subjective, depending on exactly how dry the researchers consider "dry." Nevertheless, in general it is a fairly good approximation to multiply the "dry" concentrations by

TABLE 8-5 Platinum in Human Tissue from Autopsy Cases[a]

Tissue	Platinum Concentration, μg/g (dry tissue)	
	Published data[b]	Unpublished data[c]
Stomach	—	97
Spleen	24	87
Liver	83	76
Pancreas	48	65
Lung	95	65
Ovary	—	54
Muscle, skeletal	—	54
Adrenal	—	44
Kidney	83	44
Small intestine	—	44
Bone marrow	—	39
Fat	—	33
Bladder	—	33
Gall bladder	—	33
Lymph node	60	22
Skin	—	11
Colon	—	0
Brain	—	0
Thyroid	60	—

[a] Measurements based on atomic-absorption spectrophotometry.
[b] Data from Hill et al.[199] Cause of death, acute granulocytic leukemia.
[c] Data from R. J. Speer and H. Ridgway, personal communication.

0.25, to allow direct comparison with equivalent "wet" concentrations. Even after this is done, however, the Wadley data are the highest of all those reported.

Although data in Table 8-5 are probably all much too high in absolute magnitude, the *trend* of the values may be significant. Stomach, liver, and kidney have rather high concentrations, whereas the platinum apparently does not accumulate in the brain tissue. It is interesting to note that, according to these data, fat has a much lower concentration than liver, in apparent disagreement with the conclusions of the Stewart Laboratories workers. Data on the concentration of platinum in human heart tissue have been given by Wester.[473]

In summary, it is probable that the baseline concentrations of platinum and palladium in tissues of unexposed people are both less than 3 ppb (0.003 μg/g of wet tissue). It is obvious that additional data from other parts of the United States need to be analyzed carefully, to test the validity of this statement.

Environmental Considerations 133

No baseline data for concentrations of rhodium, ruthenium, osmium, or iridium in human tissue are available. The concentrations of these metals are very likely below the sensitivity limits of current analytic instruments.

The SWRI researchers[233] have also analyzed samples of blood, urine, feces, and hair from employees at the Canadian mining facility near Sudbury and at a noble-metals refinery in New Jersey. In addition, tissue samples from autopsies of nine deceased mine employees were studied. All samples of blood, urine, feces, and hair collected from the mineworkers were below the limits of detection for both platinum and palladium. Tissue from only three of the nine autopsies had detectable platinum: fat, 4.5 ppb; lung, 3.7 ppb; and muscle, 25.0 ppb. Although samples of liver, kidney, spleen, lung, muscle, and fat from each autopsy were analyzed, these three detectable concentrations were all for *different* people, thus suggesting some sample contamination. It is therefore concluded that people who work in mining areas where platinum-group metals are extracted do *not* incorporate significant amounts of these metals into their bodies.

Similarly, blood samples collected from 61 refinery workers in New Jersey contained no measurable amounts of platinum (less than 1.4 ppb) or palladium (less than 0.4 ppb).[233] However, about 10% of the urine samples had measurable amounts of platinum (maximum observed, 2.6 μg/liter; detectable limit, 0.1 μg/liter), and over half contained measurable palladium (maximum observed, 7.4 μg/liter). No autopsy data from refinery workers are available. In view of these detectable amounts of platinum and palladium in urine, it is strongly recommended that autopsy tissue samples from deceased refinery workers be analyzed, to determine whether such exposure leads to platinum or palladium incorporation and, if so, where the metals accumulate. Such data would provide information about the probable fate of platinum-group metals in humans, if the general population became exposed to them.

Vegetation

There have been only a few attempts to determine the concentration of the platinum-group metals in vegetation.[59,63] Recently, an *indirect* method has been used to provide information about the concentration of trace metals in the environment. Researchers from New York[442] analyzed by spark-source mass spectrometry the concentrations of 47 elements in honey that had been collected near highway, industrial, and mining areas. These elements included palladium, rhodium, and ruthenium, with the middle of the range of values reported being about

9, 12, and 18 ppb, respectively. Admittedly, this technique may be subject to contamination related to the use of the containers for the honey, and the bees may tend to concentrate or magnify the elemental contaminants in the air, water, soil, and plants from which they collect nectar. Nevertheless, they forage over about 6.4 km^2 and thereby provide samples from a rather large area. This technique deserves further study, to determine its suitability for monitoring trace-metal contamination in the environment.

One note of caution regarding interpretation of the data needs to be offered: a newspaper article[468] describing the honey analysis[442] reported that all the contaminants could be traced to gasoline and lubricating-oil emission fumes, diesel additives, and catalytic converters, all used in automotive vehicles. The analysis itself was submitted to a journal for publication on *November 12, 1974,* barely a month after the first cars with catalytic converters reached the market. Furthermore, the fact that no platinum was found in the honey rules out such converters as a source of the metals, inasmuch as the converters contain platinum and palladium in a ratio of about 2.5:1. One would thus expect some platinum to be observed in the honey, if the bees had foraged in an area that contained residue from catalytic converters. This points out some dangers associated with careless interpretation of results.

Animals

Only one incomplete set of tissue data appears to have been reported for four organs of a New Zealand white rabbit.[421] According to the report, spleen, liver, and lung all contained platinum at 100 μg/g (dry weight), and kidney, at 120 μg/g (dry weight). Some of the same researchers are now measuring tissue samples from a rat (R. J. Speer and H. Ridgway, personal communication). As mentioned earlier, all these values are probably much higher than would have been obtained from the samples by either the SWRI or Stewart Laboratories researchers.

Because of the wide differences between baseline values obtained by the different laboratories, it seems obvious that comparative measurements on the *same* tissue sample need to be made by all the laboratories involved in determining baseline data.

AUTOMOBILE EMISSION CONTROL DEVICES

The newest and by far the most extensive use of platinum-group metals is in catalysts for purifying exhaust streams from automobiles.[196,210,469] Although used for many years to control emission from vehicles oper-

Environmental Considerations

ated in restricted environmental areas (fork-lift trucks, mining equipment, etc.),[1] such devices were first installed on general-purpose automobiles in the United States in October 1974. Almost all cars manufactured in the United States since then, as well as a large fraction of new imported vehicles, contain these devices. Reaction by the general public to their use has been varied and strong, with both proponents and opponents using arguments based on environmental quality, fuel economy, cost, simplicity, and reliability. Although the catalytic converters do indeed decrease emission of hydrocarbons and carbon monoxide, they introduce the possibilities of producing sulfur trioxide (or sulfuric acid after reaction with moisture), of emitting platinum-group metals and their compounds into the environment, and of creating a fire hazard through overheating. The purposes of this section are to describe the development and operation of catalytic converters and to assess their possible positive and negative impacts on the environment.

Government decisions have had an overwhelming effect on development of catalytic converters. Over 15 years ago, the California legislature enacted a law that would require all new cars sold in the state to meet specified emission standards as soon as at least two techniques could be perfected for achieving those standards.[262] The potential market being substantial, considerable effort was expended on both catalytic and noncatalytic approaches to solve the emission problem. Three catalytic devices were actually certified,[469] but they worked only with unleaded fuel. The unavailability of unleaded fuel, coupled with the ability to meet the proposed standards through minor engine modifications and carburetion recalibration to lean mixtures, stifled further development of catalytic devices for several years.

The next government decision that affected catalyst development was the U.S. Federal Clean Air Act of 1970.[349,451,454] This law required a 90% reduction in hydrocarbon and carbon monoxide emission by 1975 and in nitrogen oxide (NO_x) emission by 1976 on all new-model cars for 50,000 miles (about 80,500 km). Because some progress had already been made by manufacturers in reducing emission by about a factor of 2 since the mid-1960s, the law in fact required about a 95% decrease in emission, compared with the "dirtiest" automobiles; the technology did not exist at that time to meet these requirements on mass-produced cars. Although each of these deadlines for compliance could be legally delayed for a year (provided that it could be demonstrated that the standards could not be met after a "good-faith" effort had been exerted), this was the first time that legislation had attempted to enforce implementation of a technology that had not been perfected.

Because of the oil embargo in the 1970s, the Clean Air Act was amended by the Energy Supply and Environmental Coordination Act in 1974[350] to postpone until 1977 (or 1978, as later allowed by EPA Administrator Russell E. Train[453]) enforcement of the statutory standards and to establish some less stringent interim standards. The California Air Control Board has established even more stringent interim controls that apply to new cars sold in the state and has enforced these standards recently by levying substantial fines against some manufacturers for failing to comply. All the standards are now being reconsidered by both the executive and legislative branches of the U.S. government; it is very likely that the statutory hydrocarbon and carbon monoxide standards will be postponed at least until 1981, and there is a strong possibility that the federal NO_x standard will be permanently relaxed somewhat.[197]

When automobile emission was first being regulated, it was suggested that a tailpipe "concentration" standard be established for each pollutant. Of course, this was unrealistic, in that installation of an air pump to dilute the pollutants could produce whatever concentration was desired. It was then suggested that a standard based on "mass emission" per unit distance traveled be established, and this is the basis of all current regulations. Because the emission varies widely for a given car, depending on the mode of operation, it is essential that some standard test be established; the federal test procedure[349,451,454] now used in the United States is the "constant-volume-sampling, cold-hot start" (CVS-CH) test. With a car mounted on a chassis dynamometer and put through a complex, well-defined 11.5-mile (about 18.4-km) driving cycle (41.3 min, including a 10-min shutdown), constant-volume samples of the diluted exhaust are collected sequentially in three bags for analysis by nondispersive infrared techniques (for carbon monoxide and dioxide), flame ionization (for hydrocarbon), and chemiluminescence (for NO_x, with $x \geq 1$). Table 8–6 summarizes precontrol, present, and statutory emission, according to CVS-CH tests. It is apparent that progress has been made in decreasing emission, but even the present California standards are higher than the statutory federal requirements by a factor of about 2–5. It is now possible, through the use of catalytic converters, to meet the statutory requirements for hydrocarbon and carbon monoxide. Technology for meeting the NO_x statutory standard for 50,000 miles has not been demonstrated, and it is doubtful that it can be developed without imposing a substantial fuel-economy penalty. It is on this aspect that research needs to be concentrated.

Environmental Considerations

TABLE 8-6 Exhaust Emission from U.S. Cars

	FTP^a Emission, g/mile^b		
	Hydrocarbon	Carbon Monoxide	NO$_x$
Precontrol cars, before 1968	17.0	125.0	6.0
Present standards in 49 states (excluding California), until 1978	1.5	15.0	3.1
Present California standards	0.9	9.0	2.0
Federal Clean Air Act statutory requirements	0.41	3.4	0.4

^{a}Federal test procedure.[349,451,454]
^{b}To convert to grams per kilometer, multiply values by 0.6.

Japan has laws that required oxidation catalysts on many cars beginning in 1976 and will require NO$_x$ reduction catalysts by the end of 1978.[200]

Characteristics of Catalytic Systems

GENERAL DESCRIPTION

Before details of catalyst formulation are given, it is instructive to examine some of the system characteristics and a few of the demands that are placed on catalytic converters. A primary characteristic is transience.[469] Even when the fully warmed-up car is operated at a constant speed, the exhaust gas pulsates as the contents of each cylinder are dumped into the exhaust manifold. More importantly, as driving mode changes, the catalyst must be expected to perform under a wide variety of temperatures, space velocities (flow rates of exhaust gases through the converter), and mechanical shock conditions. It must also be resistant to poisons from the fuel, oil, or air that occasionally contact it. In general, it must be able to withstand considerable mistreatment at the hands of drivers who have not been educated to appreciate the sensitivities of catalytic materials. This is a far cry from the usual mode of operation of catalytic reactors in chemical plants or petroleum refineries, where the byword is stability.

Two different environments are required to purify the exhaust gases. For hydrocarbon and carbon monoxide control, an oxidation catalyst in a fuel-lean (oxidizing) atmosphere must be used. For NO$_x$ removal, a reduction catalyst in a fuel-rich (reducing) atmosphere is used to effect reduction by carbon monoxide, hydrogen, ammonia, or hydrocarbon. Actually, the most straightforward way of removing NO_x is by simple decomposition into nitrogen and oxygen, a reaction that is thermodynamically favorable, except at very high temperatures. However, no effective catalyst for this reaction has yet been found.

To accommodate these different environments, it was proposed very early[196] that two different catalyst beds be used in series, the first operating in a reducing atmosphere for NO_x control, and the second in an oxidizing atmosphere for hydrocarbon and carbon monoxide control, as shown in Figure 8-1. The engine would be tuned to run rich to produce a reducing (oxygen-deficient) atmosphere in the first reactor, and an air pump driven by a fan belt would inject air at point 2 to produce an oxidizing (oxygen-rich) atmosphere in the second reactor.

Although such a scheme would theoretically accomplish the objectives, it has some problems. First, if the mixture is too far on the rich side of stoichiometric in the first bed, some ammonia is formed, owing to reaction of hydrogen (produced either in the engine or via the water-gas-shift reaction in the catalyst bed) with the nitric oxide. The ammonia, not particularly harmful itself at these concentrations (there is already a measurable amount of it in the atmosphere), would be converted back into nitric oxide in the oxidizing atmosphere in the second reactor. Thus, the *net* nitric oxide conversion would not be significantly decreased. Second, operating the car in a fuel-rich mode at all times will cause a substantial fuel-economy penalty. Third, the first catalyst that will become effective (or reach its "light-off" temperature—i.e., where 50% conversion occurs) is the one for NO_x control, whereas the oxidation catalyst (further downstream) will heat more slowly. Unfortunately, it is *oxidation* activity that is more needed early in the driving cycle, beginning from a cold start, because the reducing atmosphere caused by functioning of the choke creates large amounts of carbon monoxide and hydrocarbon initially. NO_x emission

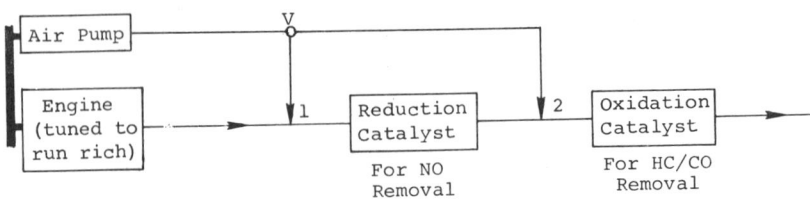

FIGURE 8-1 Dual-catalyst exhaust control scheme. V, switching valve. 1 and 2, air-injection points.

Environmental Considerations

does not become important until the engine becomes hot. To avoid this situation, it has been suggested that air be injected into the first reactor at point 1 (Figure 8-1), thus using the first reactor as an oxidation catalyst until the system becomes hot, at which time the air would be switched to point 2 for normal operation. This would sacrifice NO_x control for a short time initially, but it would increase hydrocarbon and carbon monoxide control, because a large fraction of the hydrocarbon and carbon monoxide collected during the federal test procedure comes from "Bag 1," the first of three samples collected, and includes samples of emission during the cold-start part of the driving cycle. In addition to being more complex because it involves a switching valve (V in Figure 8-1), such an approach demands extreme versatility of the first catalyst by requiring both oxidation and reduction activity. Fourth, introduction of two catalyst beds doubles the pressure drop in the system; this situation can cause accelerated engine wear and decreased performance. Finally, the excess air injected into the oxidation reactor (at point 2, Figure 8-1) will maximize the formation of sulfuric acid, because the equilibrium formation of sulfur trioxide from sulfur dioxide and air depends on the partial pressure of oxygen, as seen in the equilibrium conversion curves in Figure 8-2.

Another scheme that has been suggested is a single catalyst bed to effect removal of all three pollutants simultaneously. This three-way catalyst approach stems from the observation that, for mixtures very near stoichiometric (air:fuel ratio, about 14.7 lb of air per pound of fuel), conversion of all three pollutants is high (Figure 8-3). However, if one shifts more than 0.1 lb of air away from that point in either direction, conversion of at least one of the components decreases substantially. During normal driving, the carburetion varies over a much wider range than this. For example, during medium cruise, the ratio normally is on the lean side of stoichiometric at about 16:1, whereas it may decrease to as low as 12:1 for maximal power during rapid acceleration. Thus, it is obvious that substantial changes in engine control will be required to keep the air:fuel ratio within the ± 0.1 "window." Such control can probably be achieved only through use of an oxygen sensor and a feedback system to maintain the oxygen partial pressure exactly correct. High-temperature solid electrolytes, such as zirconium dioxide, can be used; they develop high voltages when the oxygen partial pressure approaches zero. An example of how the potential changes with the air:fuel ratio is shown in Figure 8-3. If a sensor were placed near the catalyst bed, its output voltage could be fed into a small computer that will either increase or decrease the air:fuel ratio in the carburetor as needed. Like the dual-catalyst system, the three-

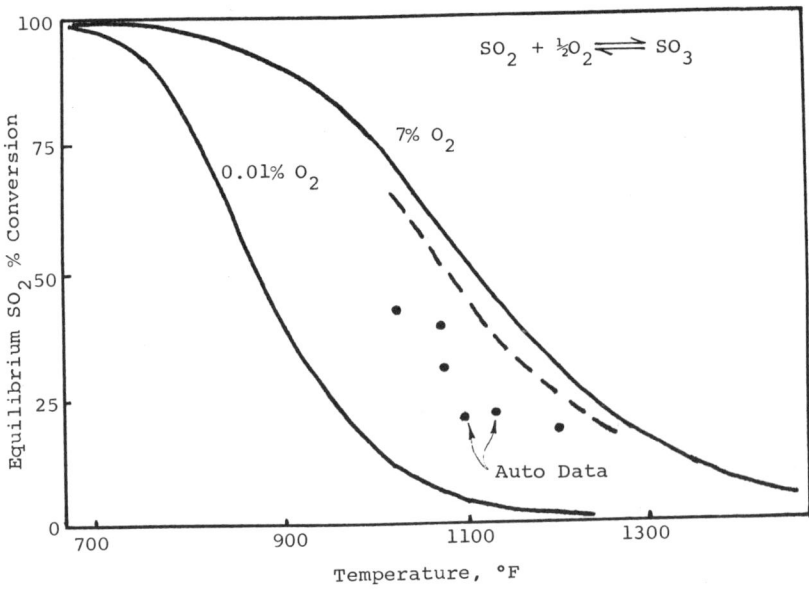

FIGURE 8-2 Equilibrium conversion of sulfur dioxide to sulfur trioxide at total pressure of 1 atm for two oxygen concentrations. Dashed curve and points refer to equilibrium and observed conversions in an automobile. Data provided by Engelhard Industries, June 10, 1974.

way approach has problems. First, the air:fuel ratio tolerance is extremely limited. Second, the sensor does not become effective until it is hot—a problem it shares with the catalyst. Third, there is a delay between the catalyst's "seeing" something and the computer's dictating of action at the carburetor. Fourth, both the catalyst window and the sensor signal may shift as the system ages, as indicated by the dashed curves in Figure 8-3. If the two curves do not shift in concert, the device may begin controlling at a point far removed from the catalyst window. On the positive side, such a system has a lower pressure drop, requires less catalyst, and minimizes the ammonia and sulfur trioxide problems that plague the dual-catalyst approach. Considerable catalytic research needs to be done to increase the width of the effective window and to stabilize the system against shifts due to aging. If the window can be widened enough, it may be possible to control the carburetion sufficiently, in a simple way, without the use of a complex oxygen sensor-feedback system.

The scheme currently being used in U.S. cars involves an oxidation catalyst to control hydrocarbon and carbon monoxide either with or

Environmental Considerations

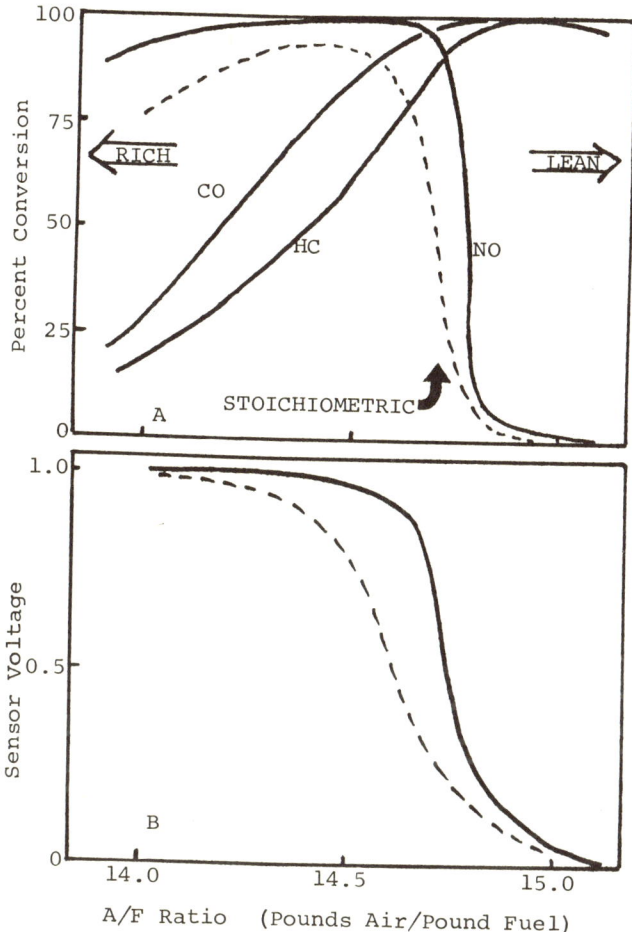

FIGURE 8-3 Performance of three-way catalyst system and oxygen sensor as a function of carburetion—i.e., air:fuel ratio. Solid curves indicate new system, dashed curves an "aged" system. Data provided by Nissan Company, May 23, 1974.

without addition of air pumps. Some degree of NO_x control is achieved with exhaust-gas recirculation (EGR), which minimizes formation of NO_x in the engine by decreasing the combustion temperature. A large fixed recirculation ratio will result in a substantial fuel-economy penalty, although a "proportional" recirculation ratio that varies with driving mode can almost eliminate the penalty. In the best case, EGR can be expected to reduce the NO_x emission to no less than 1.0 g/mile; this is far in excess of the U.S. statutory limits. Other engine forms, such as stratified-charge and diesel, also have NO_x emission in the same range, even under optimized conditions.[295] It thus appears that catalytic converters will be required, if the statutory 0.4-g/mile standard for NO_x is ultimately to be enforced.

GEOMETRY AND PHYSICAL PROPERTIES

The location and geometry of catalytic converters play quite important roles in determining their overall performance and influence the health effects that may be associated with these devices. If they are very near the exhaust manifold for rapid heating, the converters may be subject to overheating that can irreversibly damage the catalyst. Cylinder misfiring and sparkplug-wire disconnection are common malfunctions that cause overheating. In most configurations, the converters are either under the front seat or just ahead of the front floor panel.

Although at least six physical forms of catalysts have been proposed, only the first two are currently in use. The six are catalyst pellets (used by General Motors, American Motors, etc.), ceramic monoliths (placed on Ford and Chrysler products, as well as several non-American cars), layered expanded metal screens (tested by Questor[45]), coiled wire mesh (made by Gould[142]), alumina-coated wire strands (synthesized by Texaco), and metal sponges (proposed by Clyde Engineering Co.).

The "pelleted" catalysts are in the form of extrudates, spherical particles, or cylindrical pellets about 1/8 in. (about 0.3 cm) in diameter. The most popular converter is the "frying-pan" or "turtle" configuration developed by General Motors, which has a volume of 260 in.^{3} (about 4.3 liters); another version used on some smaller cars has a capacity of 160 in.^{3} (2.6 liters). The catalyst particles are held in a thin bed between two screens that are almost horizontal, but the exhaust gases enter the converter at one end, flow down through the bed, and are collected below the bottom screen and exit at the other end. The screens, which allow for particle expansion and shrinkage during heating and cooling, maintain some pressure on the catalyst bed, to minimize loose packing of particles and their collision with each other; this could cause attrition and loss of catalytic material from the exhaust system during the violent shaking sometimes encountered on rough roads. The converter has the advantageous capability of being refilled (if necessary) with fresh catalyst through a hole in its side without removal from the car. The large converter has a bed density of about 0.65 and holds a total of about 2.5 kg of catalyst.

The monolithic catalysts are single pieces of ceramic material with parallel channels running along their length. They are sometimes in what is called a "honeycomb" structure. Their length varies from about 3 to 6 in. (7.6 to 15.2 cm), and their diameter from 4 to 6 in. (10.2 to 15.2 cm), although some are oblong; and there are usually between 10 and 20 channels per inch (or between four and eight per centimeter). The channels can be triangular, square, hexagonal, or sinusoidal, depending on the manufacturer. The volume of a monolithic converter is usually one-fourth to one-half that of a particulate converter. The open structure minimizes back-pressure effects for flow rates that can approach 500 standard cubic feet (scf) per minute, or 14 m³/min (or give a space velocity—volume of gas per volume of reactor per hour—of up to 200,000/h). The monolithic catalysts suffer from being difficult (if not impossible) to replace without complete removal of the converter from the car. They are also more subject to thermal-stress cracking than are the pelleted catalysts, but attrition does not seem to be so great with monoliths as with pelleted material.

COMPOSITION OF OXIDATION CATALYSTS

The ingredients of oxidation catalysts can be divided into two parts: active catalytic material (minor component) and relatively inert support material (major component). In all cases, the active catalytic components for both oxidation and reduction are platinum-group metals or base transition metals and their oxides. These materials are dispersed either atomically or in the form of widely separated small crystallites on a high-surface-area support, usually gamma alumina.

Platinum and palladium (and sometimes rhodium) are the only active ingredients that have proved durable for application as oxidation catalysts in automobile exhaust; in most cases, a loading of about 0.06 troy oz/car is required. For pellets, the entire support is made of gamma alumina (100–200 m²/g), sometimes combined with a "stabilizer"—such as magnesium oxide, MgO; cerium oxide, CeO_2; sodium oxide, Na_2O; zinc oxide, ZnO; or titanium dioxide, TiO_2—to increase high-temperature stability and to decrease shrinkage. The noble metals are added either by a batch-impregnation technique, in which the pellets are immersed in an aqueous solution of H_2PtCl_6 or $PdCl_3$, or by a continuous-flow method, in which the preformed pellets are sprayed with a solution containing these chemicals. For best results, the platinum-group metals should be concentrated near the outer surface of the pellets, because metal buried deep inside is prevented by diffusion resistance from participating effectively in the reaction. However, some of the metal should be at least a bit below the external surface, to protect it from such poisons as lead and phosphorus that may periodically find their way into the fuel or lubricants. Wei[470] has contrasted the poisoning resistance of the "egg-yolk" catalyst with the reaction availability of the "eggshell" catalyst; a bit of both seems to be optimal, with none of the platinum-group metals deposited deep in the center of the pellets. Once impregnated, the metal salts are reduced to zerovalent metal atoms, and it is in this form that they are catalytically active. Sometimes, hydrogen sulfide is used as the reducing agent and also helps to "fix" the platinum-group-metal atoms in such a way as to prevent crystallite growth and loss of effective surface area.

The ceramic material used as a base in the monolithic oxidation catalysts is now almost exclusively cordierite, $2MgO \cdot 2Al_2O_3 \cdot 5SiO_2$, chosen mainly because of its very low thermal coefficient of expansion;[16] the low coefficient is to avoid stress cracking under the large thermal gradients that can occur during some modes of operation. Most suppliers now use extrusion processes to prepare the monoliths, although the earliest versions[349,451,454] were made from layered (or wound) corrugated cardboard-like material containing powdered ceramic that could be calcined to remove the paper binder and fix the geometry. There are about 300 channels/in.^{2} (about 46/cm^2) in the best configurations.[193] Because the ceramic has an extremely low surface area (<0.1 m^2/g), a thin "wash coat" of stabilized gamma alumina* is deposited on the insides of the channels, to serve as a support for the platinum-group metals. This wash coat is usually applied as a slurry that is forced through the channels; the excess is removed by compressed air. In some cases, the metals are impregnated from aqueous solution directly on the wash coat after it has been applied to the ceramic; in other cases, they are mixed directly with the wash-coat slurry before its application. If the wash coat is too thick or is improperly applied, it has a tendency to flake off the ceramic support and may be eliminated, with the platinum-group metals, from the converter. In any case, the final material is calcined and the crystallites "fixed" by treatment with hydrogen sulfide.

Base transition metals cannot compete with platinum-group metals as oxidation catalysts, for several reasons. First, the "light-off" characteristics (activity versus temperature) of base metals are usually much less desirable than those of the noble metals, as shown in Figure 8-4. These data were obtained by passing a stream of synthetic exhaust through a laboratory reactor that contained a small amount of catalyst. As the temperature was increased, the conversion of carbon monoxide and hydrocarbon was periodically measured. The platinum-group-metal catalysts typically yield curves that increase rapidly at a

*The exact nature of the stabilizers used is proprietary, but they probably do not involve alkaline earth oxides.

Environmental Considerations

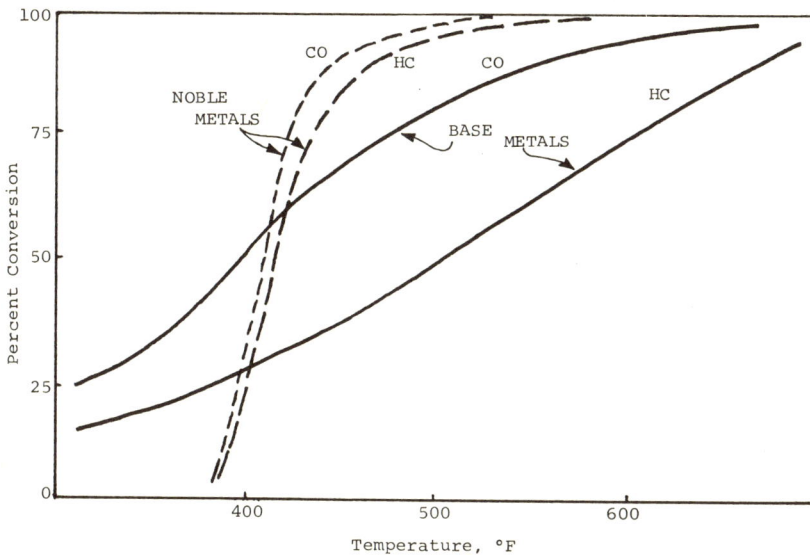

FIGURE 8-4 Activity comparison of base-metal and noble-metal oxidation catalysts in a laboratory reactor; "light-off" curves. Derived from General Motors Corporation.[158]

moderate temperature to nearly 100% conversion (a desirable characteristic), whereas the base-transition-metal catalysts require much higher temperatures to approach 100% conversion. Second, the conversion over base metals is a sensitive function of space velocity, whereas conversion over noble metals is relatively insensitive to changes in the flow rate (a desirable feature, because the flow characteristics vary so widely during different driving conditions). Third, the platinum-group-metal catalysts are not poisoned significantly by sulfur in the fuel, whereas the base metals form surface sulfates and become deactivated, until the decomposition temperature for the sulfate is reached (typically around 700° C). This is generally above the normal operating temperature for the converters. Fourth, the base metals have a tendency to react chemically with the support to form inactive spinals, such as $NiAl_2O_4$, but this can be minimized by proper choice of the combination of transition metal and support. Fifth, the base metals are less active per unit weight than the platinum-group metals by a factor of 100–1,000, and this requires a considerable quantity of the former for acceptable performance. Such large amounts can lead to a decrease in the surface area by plugging of the pores of the gamma alumina. Finally, the base metals follow essentially first-order kinetics for carbon monoxide oxidation, whereas over catalysts containing platinum the reaction order for carbon monoxide is *inverse* first-order in carbon monoxide concentration. This means that, as the partial pressure of carbon monoxide becomes smaller over platinum catalysts, the oxidation rate increases; this makes such catalysts exceedingly effective for removing small amounts of that pollutant.

COMPOSITION OF REDUCTION CATALYSTS

There are two catalytic routes for removal of nitric oxide: direct decomposition and chemical reduction. No catalyst has been found that will decompose nitric oxide effectively in an oxidizing atmosphere.[18] This is primarily because the released oxygen remains strongly attached to the surface and poisons the sites for further catalytic activity. Therefore, most research on nitric oxide removal has centered on finding active and stable catalysts for reduction by carbon monoxide, hydrogen, and hydrocarbon. The problem is more complex for oxidation, because a selectivity factor is involved. Ammonia can be formed if the mixture is too rich, and in some cases nitrous oxide is formed in significant amounts. It has been suggested[244] that one of the primary routes by which nitric oxide is reduced is through ammonia as an intermediate, which is illustrated by the imbalanced reactions a and b.

$$NO + H_2 \xrightarrow{a} NH_3 \xrightarrow{b} N_2 + H_2$$

Copper, platinum, and palladium are good catalysts for reaction a, but they do not catalyze the ammonia decomposition effectively. However, nickel is a good catalyst for the ammonia decomposition (reaction b). Thus, a combination of the noble metal with nickel, or a mixture of copper and nickel (e.g., the alloy Monel, which has a copper:nickel ratio of 30:70), is a good catalyst for this reaction. Considerable work has been done by Amoco, Exxon, and Gould on the Monel systems, and similar material is still being seriously considered as a catalyst for nitric oxide reduction in cars. The main problem stems from its susceptibility to physical deterioration during the cycling from oxidizing to reducing atmospheres that occurs in normal driving. It does, however, have the advantage that it can be used either as self-supporting homogeneous alloy chips (or wires) or in a form that can be supported on ceramic or metal supports.

By far the most effective catalyst for nitric oxide reduction is ruthenium.[245] It is so active that only a very small amount of it (perhaps as little as one-tenth the amount of platinum in oxidation catalysts) is required, and it does not form significant amounts of ammonia, even under extremely rich conditions. Unfortunately, the metal can form the volatile ruthenium trioxide and tetroxide, RuO_3 and RuO_4, under oxidizing conditions, and this leads to its slow removal from the catalytic converter. The ruthenium can be "stabilized" through formation of ruthenates, such as lanthanum ruthenate, $LaRuO_3$; barium ruthenate, $BaRuO_3$; and magnesium ruthenate, $MgRuO_3$. The latter is formed by impregnation of magnesium oxide, MgO, with an aqueous solution of ruthenium trichloride, $RuCl_3$.[403,404] Although there is some dispute about this point, it appears that even the "stabilization" techniques are not sufficient to guarantee that ruthenium will not be depleted from the catalyst in actual operation. It is therefore doubtful that ruthenium can be used for this purpose, inasmuch as its oxides are quite toxic.

One of the earliest catalysts used for nitric oxide reduction was copper chromite.[379] Although this material has good initial activity, it is subject to poisoning and physical deterioration. More realistically, mixtures of copper, cobalt, nickel, and chromium have been electrodeposited onto thin metal foils by Gould to form active expanded metal mesh catalysts. This material is a good reduction catalyst, but it also suffers physical deterioration (called "green rot," a concentration of chromium at the grain boundaries that leads to flaking of the electrodeposited material) when subjected to rapid oxidation–reduction cycling. It should be noted that both nickel[297] and Cr^{6+} (as chromate)[296] are regarded as potential carcinogens. To avoid such cycling, Gould has suggested installing a third catalyst—another oxidation bed, called a "getter," just upstream of the reduction catalyst, to remove excess oxygen through reaction with carbon monoxide and hydrocarbon. Addition of the third catalyst allows the engine to be operated very near the stoichiometric point (slightly reducing)—a factor that minimizes the fuel-economy penalty associated with rich operation.

All the catalysts thus far described require a reducing atmosphere (no excess oxygen). However, workers at Exxon[120] recently reported that iridium supported on alumina, Al_2O_3, can selectively catalyze the nitric oxide–carbon monoxide reduction reaction, even in the presence of excess oxygen. Much more work will be required, to determine whether this effect can be maintained in a system that is more realistic than the laboratory reactor and synthetic gas mixture used in their tests.

COMPOSITION OF THREE-WAY CATALYSTS

Three-way catalyst systems are designed to operate near the stoichiometric point, so formation of ammonia is not usually a problem. Furthermore, absence of excess oxygen minimizes the formation of sulfur trioxide. Finally, the possibility of using only a single catalyst bed makes this system extremely attractive. The very narrow range of air:fuel ratio for effective performance is the only factor that detracts from this approach, but this turns out to be a major disadvantage. The ordinary platinum–palladium oxidation catalysts simply do not have sufficiently wide windows to be prime candidates. However, catalysts containing rhodium have a 90% conversion window (see Figure 8–3) of about ±0.3 air:fuel units, and additional research is under way to stretch this window even more. Perhaps addition of some iridium will help to extend the window on the oxidizing side. If the window can be widened, it may be possible to use such a system *without* an oxygen sensor and feedback loop by properly designing the carburetor. There is optimism that this can be achieved, and it would greatly simplify emission control systems, although the availability of rhodium would probably be a significant problem.

Another possible three-way catalyst was recently patented by an inventor at Du Pont[251] who claims that his material is active for both oxidation and reduction and, more important, is *not* poisoned by lead in the fuel! It is a perovskite type of AB_2O_3 material, with ruthenium or platinum substituted into 1–20% of the B sites. The remainder of the B sites contain cobalt, and the A sites are occupied by a mixture of lanthanide ions. The catalyst is being tested in automobiles. One characteristic is that it is less active than the oxidation catalysts and thus operates at considerably higher temperatures than are currently being used.

Beneficial Effects of Catalytic Converters

Since their installation on most U.S. cars beginning in the 1975 model year, exhaust catalysts have been used to reduce the emission of carbon monoxide and hydrocarbons from light-duty vehicles. Relative to the emission of these substances in 1968, carbon monoxide and hydrocarbon emission was reduced (in 1975) by 90% in California and by 83% in the other 49 states.

The reduction in ambient carbon monoxide will increase the oxygen-carrying ability of the blood, which should substantially benefit vulnerable subgroups of the population, including people with vascular disease or severe anemia.[284] There may well be measurable improvements in the ability of people to perform tasks that require integration of the sensory, judgmental, and motor functions of the central nervous system.[284]

The reduction in hydrocarbon emission will reduce the concentra-

Environmental Considerations

tions of photochemical oxidants in ambient air, and this should result in fewer days that have oxidant concentrations high enough to irritate the eyes and the respiratory tract.[284] This reduction in oxidants should also enhance the functioning of the physiologic mechanisms that protect and clean the respiratory system and should reduce the frequency of asthmatic attacks.[284]

In addition to the benefits associated with reduction in emission constituents regulated by the Clean Air Act, substantial reductions in several types of unregulated emission are effected by the use of catalytic converters, with resulting benefits to human health. Because nonleaded fuel is required, lead emission into the atmosphere will begin to decrease, and this should reduce community lead burdens and prevent further buildup of lead in soils and other components of the ecosystem.[284] There should also be a reduction in emission of fine particulate matter consisting of a core of lead compounds and a shell of complex organic matter[9]—including compounds known to be carcinogenic—with a concurrent decrease in the risk of respiratory malignancy. Inasmuch as the halide scavengers used with tetraethyl lead will also be absent, the emission of particulate and gaseous halides will be reduced.

Vehicular exhaust contains polycyclic aromatic hydrocarbons, including such carcinogens as benzo[a]pyrene, benz[a]anthracene, chrysene, and dibenzopyrenes.[172,284,308] These compounds are destroyed with greater than 95% efficiency with catalytic converters.[308] Alkylbenzene pollutants, such as m-xylene and 1,3,5-trimethylbenzene, can form compounds that cause appreciable eye irritation through nitrogen oxide-induced photooxidation. Conversion of 91–99% is obtained for these compounds through the use of catalytic converters.[308] In addition, such compounds as phenols and aldehydes, which are among the unhealthy products of incomplete fuel combustion in the engine, are effectively eliminated by catalytic converters.[172]

Probably the most dramatic demonstration of the health benefits likely to be derived from the use of catalytic converters is the recent study[223] in which several species of mammals were exposed continuously to diluted exhaust (air : exhaust ratio, 10 : 1) from an engine operated on nonleaded fuel. In some cases, the diluted exhaust was exposed to "synthetic sunlight"; in half the experiments, the exhaust was first passed through a catalytic converter. In the absence of the catalyst, infant rat mortality was 77% within 7 days with the diluted exhaust and 100% in 5 days with exhaust that had been irradiated. When the catalyst was used, mortality was zero in all cases, including one run in which the fuel contained 0.1 wt% sulfur (about 3 times the current national concentration). Similarly, with adult rats and hamsters, significant pathologic and hematologic effects were observed only in animals that had been exposed to exhaust not subjected to catalytic conversion. These pathologic effects were not due to the high concentration of carbon monoxide in the exhaust (500 ppm after dilution), as shown by a control experiment with 500-ppm carbon monoxide in clean air. No platinum metal was found in the tissue of animals that had been exposed to the catalytically converted exhaust. Earlier animal studies with non-catalyst-treated exhaust have shown that it can produce tumors,[433] weight loss,[433] decreased pregnancy rate,[214,256] decreased survival rate,[214,256] and increased susceptibility to pulmonary infection.[214] Catalytic converters will have a positive influence on reducing all these adverse effects that have been attributed (at least in part) to uncontrolled automobile exhaust.

Another distinct advantage of catalytic converters is their contribution to improvement in fuel economy,[28] as shown in Figure 8–5. Note that the dramatic fuel-economy increase coincided exactly with introduction of the catalytic converters that effectively decouple[155] emis-

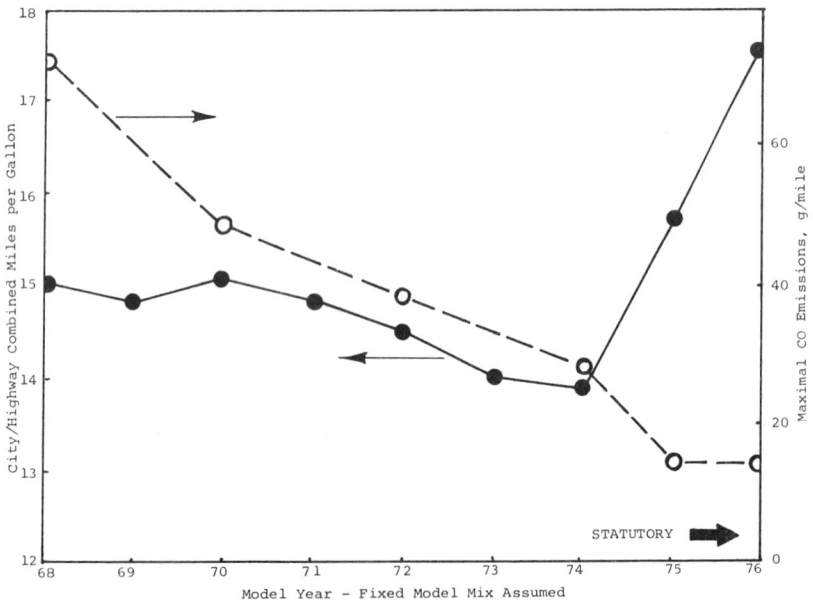

FIGURE 8–5 Fuel economy (solid line) and carbon monoxide emission (dashed line) for a fixed model mix of cars from 1968 to 1975. Catalytic converters were introduced into the 1975 model-year cars. Reprinted with permission from Austin et al.[28]

Environmental Considerations

sion control from engine performance and allow the latter to be optimized.

Potential Hazards of Catalytic Converters

Although catalytic converters have been remarkably effective in reducing hydrocarbon and carbon monoxide emission and contributing to improved fuel economy, they have introduced some health-related problems that are considered in this section. The more important ones are emission of sulfuric acid mists and possibly emission of toxic noble-metal particulate matter in the respirable size range. In addition, the tendency for the converters to become extremely hot during some engine malfunctions raises the possibility of fire inside the vehicle or in dried grass under the car.

THE SULFURIC ACID PROBLEM

All gasoline contains sulfur chemically bound in some of the fuel molecules; the amount varies according to the origin of the crude, the processing techniques, and the blending policy.[208] For leaded gasoline, the national average sulfur content is about 0.03 wt% (300 ppm), although it may vary considerably from one location to another; the sulfur content tends to be higher on the West Coast than in the rest of the country. Nonleaded gasoline usually has a lower sulfur content (around 0.01 wt%), owing to the blending of high-octane refinery streams that require extremely low-sulfur feedstocks (a few parts per million) in order to function properly—e.g., streams from the reforming and alkylation units.[208] Another low-sulfur blending stream often used contains n-butane. The only significant source of sulfur in nonleaded gasoline is the catalytic cracker stream, which may contain as much as 0.1 wt% sulfur. There are, however, ways of decreasing the sulfur in this stream by desulfurizing the feed stream to the catalytic cracker, although this entails substantially increased capital costs.

Essentially all the sulfur in the fuel comes out of automobile engines as sulfur dioxide; this source represents approximately 1% of all the sulfur emitted into the atmosphere in the United States as a result of man's activities.[134] This sulfur dioxide is slowly (over a period of hours to days, depending on the atmospheric conditions,[291-294]) oxidized to sulfur trioxide, which reacts with moisture and atmospheric bases—e.g., ammonia or metal oxides—to form sulfuric acid and airborne sulfate salts. These are eventually purged from the atmosphere by rain or through other mechanisms.

In the catalytic converters, a part of the sulfur dioxide is oxidized to sulfur trioxide and comes out of the tailpipe as aerosol particles 0.1–1.0 μm in diameter. That sulfur dioxide oxidation occurs is not surprising, inasmuch as platinum was the first commercial catalyst used for sulfuric acid manufacture (see Chapter 3). The catalytic converters do not alter either the total amount of sulfur emitted to the environment from automobiles or its ultimate disposition, but they do change the chemical form in which it appears in the vicinity of roadways. Because sulfur trioxide and sulfuric acid are considered to be more physiologically harmful than sulfur dioxide, their accumulation near roadways could represent a potential health hazard. Indeed, mathematical models[455] have predicted that, under "worst-case" conditions (crowded 10-lane freeway with adverse meteorology, such as zero wind velocity), the sulfuric acid concentration could reach 600 μg/m^3 10 ft (3 m) above the center of the roadway. Considering that the threshold[143] for irritability in sensitive people based on 24-h exposure appears to be around 10 μg/m^3, such a large value (*if* it ever occurred) would certainly pose a threat to the health of people who live near freeways or spend considerable time traveling on the roads, visiting crowded shopping centers, or working in walled city-street canyons.

The maximal concentration predicted by mathematical modeling has been the subject of intense debate. Proponents have argued that it may be off by no more than a factor of 2 in *either* direction, which would mean a range of predicted "worst-case" values of 300–1,200 μg/m^3. Opponents[124] have suggested that such modeling is highly unrealistic and that several important factors that would tend to *decrease* the concentration are not being considered. These are illustrated in Figure 8-6 for an altitude of 10 ft (3 m) at various distances from the freeway. First, the original model assumes a *single*-line source (actually, a series of closely spaced points, e.g., exhaust stacks, for which plume equations are well known) at the center of the roadway. Curve A is the predicted stationary-state sulfuric acid concentration profile. The situation is more accurately described by 10-line sources (one for each lane); the main effect of this assumption is to decrease the maximal concentration at the center of the roadway, as illustrated by curve B. Second, wind plays an extremely important role in determining the concentration profile.[124] For example, curve B becomes curve C when the wind velocity is increased from zero to 0.2 mph (0.3 km/h) at right angles (left to right) to the road. This drops the maximal concentration from about 500 to about 80 μg/m^3. At a wind velocity of 1.0 mph (1.6 km/h), the maximal concentration drops to about 25 μg/m^3 10 ft (3 m) above the downwind edge of the road. Third, the original model does

Environmental Considerations

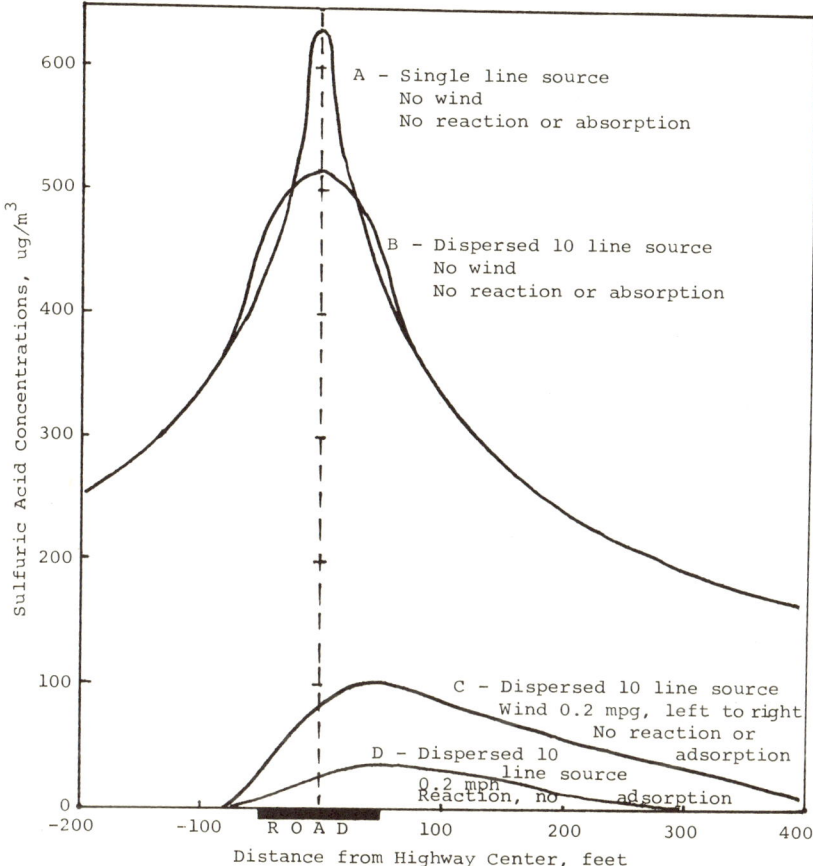

FIGURE 8-6 Effect of source distribution, wind, and atmospheric reaction on the "worst-case" maximal sulfuric acid concentration around a crowded 10-lane freeway. Reprinted with permission from Doelp et al.[124]

not account for atmospheric reactions (e.g., ammonia or metallic oxide particulate matter) that form less toxic compounds, nor does it allow for irreversible adsorption of the sulfur trioxide and sulfuric acid on surfaces that are in the vicinity of the roadway.[124] It is estimated that urban ambient air normally contains particulate matter at a concentration of $60-200 \mu g/m^3$ and ammonia at about $2-200 \mu g/m^3$. These values are large enough to allow significant neutralization of the acidic species. With modest rate parameters and reasonable values for collision between an aerosol and either ammonia or particulate matter, one can calculate curve D for a wind velocity of 0.2 mph (0.3 km/h). Of course, the physiologic activity of the sulfates themselves is not zero, although they are usually less toxic than sulfur trioxide and sulfuric acid. Adsorption on surfaces would reduce the concentration profile even further, but calculations indicate that adsorption has only small effects on the concentration curves.

Some of the characteristics that were used in these calculations are:

- fuel consumption, 15 miles/gal (24 km/gal, 6.3 km/liter)
- nominal highway speed, 55 mph (88.5 km/h)
- gasoline density, 6.5 lb/gal (779 kg/m^3)
- sulfur in fuel, 150 ppm (0.015 wt%)
- conversion of sulfur to sulfuric acid, 30% (0.027 g sulfuric acid emitted per mile, or 0.017 g/km)
- catalyst-equipped cars, 100%
- 10-lane freeway
- 0.5 car passing a point per second in each lane

This has not been intended as a thorough treatment of the subject, but it should suffice to indicate some of the enormous parametric sensitivities involved. Until more reliable experimental values can be obtained for some of the critical quantities, it appears that modeling is not very useful, in that one can obtain almost any number that will support one's own view of catalytic converters.

Apart from the stationary-state atmospheric concentrations, several factors affect the amount of sulfuric acid actually emitted by cars. First, at relatively low converter temperatures, the sulfur trioxide can react with the surface of the alumina support and be retained, or "stored," in the converter as aluminum sulfate, $Al_2(SO_4)_3$. The surface area available in a typical converter packed with pellets is around 100 acres (0.4 km^2), so the storage capacity is considerable. When the decomposition temperature of the sulfate (about 770° C) is reached, all the sulfur trioxide that has been collected is "dumped." This storage effect is more pronounced with pelleted catalysts than with monoliths, owing to the larger amounts of alumina in the former. Such an effect makes measurements of the sulfur trioxide and sulfuric acid emitted from the tailpipe heavily dependent on the driving cycle and previous history of the converter; many operational modes do not cause the decomposition temperature to be reached. Another factor is the amount of oxygen present in the converter. Figure 8-2 shows the effect of partial pressure of oxygen on the equilibrium fractional conversion of sulfur dioxide as a function of temperature. Obviously, the more oxygen present, the greater the amount of sulfur trioxide that can be formed at a given temperature. The data points collected from an

Environmental Considerations

actual automobile test indicate that equilibrium conversion (compare with dashed curve) is not quite achieved in practice. Systems that have essentially no excess oxygen (e.g., the three-way catalyst system) are thermodynamically constrained from producing much sulfur trioxide and sulfuric acid. Cars running just on the lean side of stoichiometric and without air pumps will, of course, be favored (i.e., will emit smaller amounts), although they will be less effective for the oxidation of hydrocarbon and carbon monoxide, because of the limited supply of oxygen.[392]

Another problem related to sulfur occurs when the catalyst is operated under reducing conditions. When the atmosphere is oxidizing, sulfur trioxide and sulfuric acid are formed; under stoichiometric conditions, the sulfur remains as sulfur dioxide; but, under reducing conditions, some of the sulfur is converted into hydrogen sulfide, a far more toxic substance than the sulfur oxides. This phenomenon accounts for the "rotten egg" odor occasionally reported by an owner of a new car when the car is started. When the choke is activated, a rich mixture is produced. Fortunately, the catalyst is also cold under these conditions and is not active. However, if it reaches its "light-off" temperature before the choke is released, hydrogen sulfide can be produced. Any system designed to operate very close to the stoichiometric point (e.g., a three-way system) runs the risk of forming either sulfur trioxide or hydrogen sulfide if the carburetion shifts just slightly to the lean or rich side of stoichiometric. Frequently, when the engine is turned off, an odor due to hydrogen sulfide is observed.

Finally, attempts have been made to develop effective oxidation catalysts for hydrocarbon and carbon monoxide that will not oxidize sulfur dioxide. Although there have been reports that catalysts containing rhodium partially satisfy this goal, the Subcommittee on Platinum-Group Metals doubts that such catalysts can be developed.

General Motors has conducted a series of tests to determine experimentally the concentration profiles of sulfuric acid–sulfates that could build up around a freeway.[74] A mixture of 352 cars—including General Motors, Ford, Chrysler, and American Motors vehicles—were run under carefully controlled driving cycles on the General Motors proving grounds near Detroit during October 1975. All the cars had catalytic converters, and the nonleaded fuel contained about 0.03 wt% sulfur. During the test, samples of airborne particulate matter were collected at various altitudes and at several distances from the test track; samples were also taken inside some of the cars. The road included four lanes, two in each direction. An inert tracer compound, silicon hexafluoride, was also emitted at a known rate from some test cars; measurements of this compound served as a reference against which the sulfate data were compared. Background sulfate concentrations were also measured and subtracted from the test data.

The results of the test can be summarized as follows:

- The average sulfate emission rate was 0.037 g/mile (0.023 g/km)—measured as sulfuric acid—when the vehicles were fully warm. Because of the storage effect, the observed rate was lower than this at the start of each test.
- The maximal sulfate concentrations were observed at an altitude of 0.5 m in stations at the edge of the road (on the downwind side) or in the median, the average maximal increase above background being 5.2 $\mu g/m^3$ for all runs.
- The average maximal increase in sulfate concentration inside the cars was 4.0 $\mu g/m^3$; opening or closing windows made very little difference in this value.
- Adsorption of sulfate on surfaces appeared to play a minimal role as an atmospheric removal mechanism.
- The vertical dispersion of sulfate was much greater than had been predicted by the models. This may have been because of failure to account adequately for the effects of turbulence created by the cars or for the thermal updraft due to emission of hot exhaust gases. Figure 8-7 is a contour map of the lines of constant sulfate concentration observed in one of the runs.

The values obtained in the sulfate-dispersion experiment can probably be multiplied by a factor of about 7, to allow comparison with the "worst-case" predictions of the models shown in Figure 8-6. Obviously, the observed values are far below those predicted by the original models.

Finally, several stations have been set up to monitor continuously the sulfate concentrations near freeways, especially in California. With 15-20% of the cars on the road now having catalytic converters, most of these stations have found only very slight (if any) increases in meaured sulfate concentrations.

It thus appears that atmospheric sulfuric acid accumulation due to catalytic converters will *not* pose a serious threat to human health. However, it is recommended that the concentrations around typical freeways continue to be carefully monitored, to make certain that they do not become excessive. If they begin to increase toward a potentially dangerous point (which is unlikely), one possible solution would be to limit the amount of sulfur allowed in fuels. However, there appears to be no justification for such action at present.

Environmental Considerations

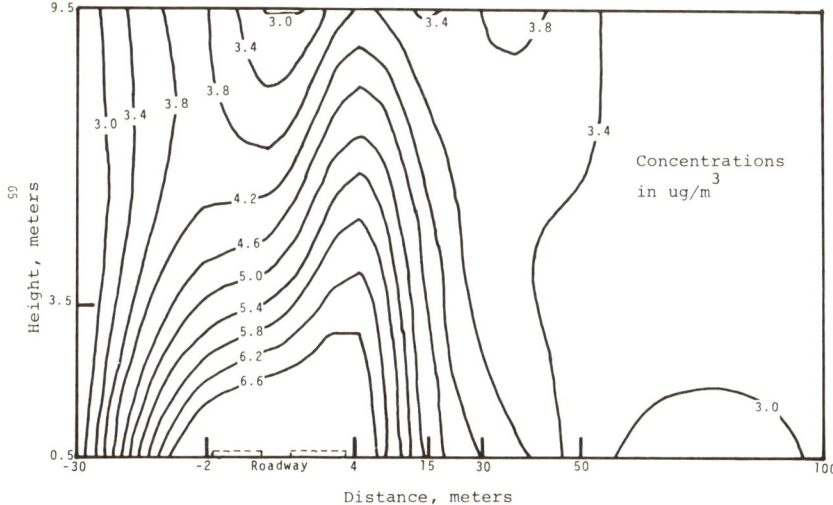

FIGURE 8-7 Contour map showing total sulfate (including background) reached at various heights above and distances from a four-lane freeway during one of the General Motors sulfate-dispersion experiments. Reprinted with permission from Cadle et al.[74]

NOBLE-METAL EMISSION

The active platinum metals are present in automobile catalysts as small crystallites widely separated on the high-surface-area alumina support. It is possible that, through attrition or malfunction modes that lead to overheating of the catalyst, pieces of the support or active ingredient can be emitted from the tailpipe. Attrition is greater for pelleted catalysts than for monolithic configurations, owing to the relative grinding motion of particles in the first case. Emission of catalyst particulate matter can be divided into two categories: large chunks that are usually collected in the exhaust muffler downstream from the converter and very fine particles that come out of the tailpipe. With most pelleted catalysts, the first category predominates; such material does not enter the environment. Attempts to collect this material on fine filters and analyze it have indicated (R. F. Hill, personal communication) that this mode of loss amounts to 1-3 μg of platinum metal per mile (0.6-1.9 μg/km), which means that less than 10% of the metal is lost during 50,000 miles (80,500 km) of operation.

The Subcommittee arranged to have samples of the two types of particulate matter, obtained from General Motors, analyzed by the Southwest Research Institute (D. E. Johnson, personal communication). The large particles obtained from sweepings of the downstream exhaust system contained platinum and palladium at 3.58 and 2.45 μg/g, respectively. One sample of the fine material contained no measurable platinum or palladium; a second sample contained platinum at 0.107 μg/g and no detectable palladium.

With an assumed emission rate of 3 μg/mile (1.9 μg/km) and with the assumption that these small particles behave in the same way as sulfuric acid aerosol particles, one can calculate a "worst-case" concentration maximum that might accumulate in the atmosphere. The maximal possible concentration under these conditions is about 0.06 μg/m^3 (compared with the 600 μg/m^3 predicted for sulfuric acid; see Figure 8-6), which is only 3% of the exposure maximum now in effect for the *soluble platinum salts*. As discussed earlier, the actual value would be less than this by at least a factor of 10 under any kind of realistic meteorologic conditions.

Perhaps the most important question concerns the chemical nature of the emitted platinum. As shown in Chapter 7, platinum metal and platinum oxides (the most probable chemical forms emitted from the converters) have little or no physiologic activity when inhaled. There has never been any evidence of physiologically active complexes of platinum among collected exhaust samples. The most definitive work in this subject was conducted by General Motors researchers who used radioactively labeled platinum to enhance analytic sensitivity (R. F. Hill, personal communication). The radioactive metal was produced by irradiating the catalytic material with neutrons. Several screens of different mesh were used to collect samples of the emitted material. About 80% was collected on a 120-mesh screen (which could trap particles greater than 125 μm in diameter), and 20% was collected on a glass-fiber filter. The platinum loss varied from 1 to 3 μg/mile (0.6 to 1.9 μg/km), depending on the driving cycle used. In all cases, the platinum:support ratio in the collected material was about the same as that in the original catalyst; this indicated that the platinum metal was not selectively lost. To obtain information about the chemical nature of the platinum, attempts were made to dissolve the material in different solvents. No radioactive platinum could be detected in the aqueous fractions; thus, less than 1% (the lower limit of sensitivity) of the platinum was presented in the form of soluble salts.

It is interesting to note that, in the General Motors sulfate-dispersion test described earlier, no platinum metals could be found in any collected samples.

It is concluded that emission of noble metals from catalytic converters does not pose a significant threat to human health through inhalation of the exhausted material.

Environmental Considerations

As noted above, most of the emitted material leaves the converters as relatively large particles (>100 μm). This material would be deposited along the roadside, where it could accumulate and be subjected to action by the environment. Wood[487] has suggested that platinum might be solubilized through methylation by specific microorganisms in much the same way that mercury is methylated. If this occurs, it could conceivably become incorporated into the food chain and become concentrated in its progress up the biologic ladder to humans. Taylor and Hanna[435] have reported recently that compounds containing platinum(IV), such as potassium hexachloroplatinate, can be methylated, but there is no evidence that platinum metal can be methylated unless it is first converted to platinum(IV). Furthermore, there is no evidence that compounds containing platinum(IV) are emitted from the converters or that they are formed in the environment after being deposited along the roadside. Even if methylation or some other form of solubilization occurs, the very small quantities of the materials involved make it remote that this could cause a significant problem in the foreseeable future. For example, to determine the extent of this possible problem, one can calculate a concentration of platinum that would be obtained after 10 years, assuming heavy vehicular traffic (50,000 cars/day, all with catalytic converters), typical emission of noble metals (3 μg/mile, or 1.875 μg/km), and accumulation in the topsoil (uniformly distributed 12 in., or 30.5 cm, deep over a width of 300 ft, or 91.4 m, with a soil density of 1.5 g/cm^3). In a test area 1 mile, or 1.6 km, long, 548 g of the noble metal would be emitted in about 7×10^{10} g of soil. This would result in a concentration of 0.008 ppm, which is at least 2 orders of magnitude *less* than the concentration in ores that are considered to be economically minable. Under these assumptions, more than 1,000 years would be required to accumulate enough material to be considered significant. Thus, it is concluded that this should not pose a significant problem. Nevertheless, it is recommended that the topsoil near a few heavily traveled freeways be monitored periodically for accumulation of platinum-group metals.

OTHER NONREGULATED EMISSION

Workers at the Bell Telephone Laboratories[460] have shown recently that it is possible to form hydrogen cyanide by passing synthetic exhaust streams over platinum catalysts in laboratory reactors under reducing conditions. However, they showed that the presence of sulfur and water in the stream completely poisons this reaction. Because both are normally present, it is highly unlikely that detectable amounts of hydrogen cyanide can be found under ordinary conditions.

TABLE 8–7 Temperatures of Exhaust Systems with and without Catalytic Converters[a]

Temperature, °F (°C)

Driving Mode	Noncatalyst System			Catalyst System		
	Exhaust Manifold	Exhaust-Pipe Converter Location	Muffler	Exhaust Manifold	Converter	Muffler
Idle	820 (438)	482 (250)	488 (253)	802 (428)	528 (276)	490 (254)
Idle and soak	805 (429)	428 (220)	439 (226)	773 (412)	543 (284)	492 (256)
City traffic	712 (378)	389 (198)	331 (166)	675 (357)	314 (157)	313 (156)
City traffic, soak	678 (359)	389 (198)	349 (176)	661 (349)	345 (174)	325 (163)
55 mph	858 (459)	371 (188)	307 (153)	822 (439)	227 (108)	292 (144)
55 mph and soak	818 (433)	374 (190)	381 (194)	788 (420)	495 (257)	389 (198)
70 mph	966 (519)	369 (187)	324 (162)	943 (506)	256 (124)	325 (163)
70 mph and soak	903 (484)	474 (246)	446 (230)	883 (473)	556 (291)	451 (233)
Two plugs not firing:						
Idle	519 (271)	528 (276)	395 (202)	538 (281)	875 (468)	735 (391)
Idle and soak	513 (267)	523 (273)	390 (199)	534 (279)	860 (460)	709 (376)

[a] Data from General Motors.[159] Additional data from V. Haensel (personal communication): Temperature to ignite pine needles: in 1 min, 760° F (404° C); in 4 min, 660° F (349° C). Temperature to ignite dried grass: in 1 min, 840° F (449° C); in 4 min, 750° F (399° C). Average skin temperatures of converters: normal operation, 550° F (288° C); 25% engine misfire after 25-mph cruise, 740° F (393° C).

Environmental Considerations 161

As mentioned early in this chapter, ammonia is formed over many proposed NO_x catalysts when the carburetion is set to give a strongly reducing atmosphere. However, in the dual-bed configuration (see Figure 8-1), the ammonia formed in the reducing reactor would be converted back into nitric oxide in the oxidizing reactor; hence, ammonia would not be emitted from the overall system.

OVERHEATING

The two most frequent modes of catalyst deactivation are poisoning by lead in the fuel and engine malfunction that causes overheating. The first mode is somewhat reversible, in that most oxidation catalysts can tolerate about one tankful of leaded gasoline in 10 without severe damage.[6] However, if the temperature becomes high enough to melt the catalyst or its container, the damage is irreversible.

The operating temperature in the oxidizing catalytic converters is a sensitive diagnostic measure of engine performance. One person has suggested that the converters serve well as "rectal thermometers." With misfiring spark plugs, missing wires, or severe carburetion maladjustment, the unburned fuel is burned in the catalytic converter, where neither cooling water nor mechanical energy can remove the heat. In cars without catalytic converters, this extra oxidation process does not occur, and unburned fuel is vented into the atmosphere. Table 8-7 compares the temperatures observed in cars with and without catalysts and operated under various malfunction modes. Note that the exhaust temperature actually *decreases* with the degree of malfunction with catalyst cars.

There has been considerable discussion recently about the fire hazards associated with these devices. Catalyst cars have been banned from many petroleum refineries, and consideration has been given to banning them from some national parks. Without question, the possibility of fire does exist, and grass fires have reportedly been caused by cars with catalytic converters. However, what is not usually appreciated is that *noncatalyst* cars, particularly those produced since 1972, also have very hot exhaust and can themselves start grass fires. This was dramatically demonstrated in a test in California planned by Los Angeles County Supervisor Kenneth Hahn.[489] Two American Motors Hornets, a 1974 model without a catalytic converter and a 1975 model with a converter, were allowed to idle for 30 min and were then driven onto a patch of dried grass. With the engines still running, the catalyst-equipped car ignited the grass in 55 s, but the noncatalyst car caused a fire in only 35 s! Another test, in which two spark-plug wires were removed from the catalyst car, caused the grass to ignite after 14 s; other tests to confirm this result were unsuccessful.

The Department of Transportation (DOT) has been keeping a file (Docket 75-13)* on all reported cases of overheating. As of March 1975, General Motors had reported 327 cases of overheating out of 2.2 million cars sold; of these, almost half were discovered by dealers before the cars were turned over to purchasers. Ford had reported 78 cases out of 1.2 million cars sold during the same period. The vast majority of these problems were traced to the ignition system's being improperly assembled or not working properly. A few cases were attributed to improper application or the wrong kind of undercoating material directly onto the converters. In several cases, floor carpets have been scorched; but fewer than a dozen actual fires have been reported. According to the DOT file, there have been no reports of human injury caused directly or indirectly by problems associated with catalytic converters.

The number of problems reported is remarkably low, considering the newness of the technology, and reflects a highly successful introduction into the market. This is reflected in a statement by the EPA administrator[453] in his decision to postpone enforcement of the more stringent 1977 standards for a year:

In many ways, catalysts have performed far better than some predicted when the 1975 interim standards were first established two years ago. Contrary to many predictions, both the production of catalysts and their installation on automobiles is proceeding without difficulty.

MATERIAL SUPPLY

As noted earlier, the amount of platinum required in catalytic converters each year is approximately equal to the total amount of platinum imported into the United States each year for the last several years. This means a doubling of our requirements for this strategic material. Although world production (mainly from South African mines) has been expanded to meet this demand, the total dependence on foreign sources is undesirable. It has been suggested that the noble metals might be recycled to a significant degree.[248] However, the small amounts of noble metal in the converters (about 0.06 troy oz, or 1.87 g, worth about $10 at current prices) and the low concentration (about 0.05 wt%) make it unlikely that this will ever be accomplished. It has been

---

*The National Highway Traffic Safety Administration Docket 75-13 regarding the risk to the public was closed February 24, 1977, without any rule-making (Federal Register 42:12284-12285, 1977).

Environmental Considerations

argued that refinery catalysts containing platinum can be almost 100% reclaimed and that the same thing could be achieved with automobile catalysts. There are two significant differences: large quantities of the reforming catalysts are present in a single location, and the concentration in these catalysts (about 0.3 wt%) is several times higher than it is in the automobile catalysts. The logistics of removing the converters, emptying the contents into containers, and shipping them back to the refinery will hardly be economical, unless the price of platinum becomes dramatically higher. However, favoring recycling is the fact that the stainless-steel converter material in which the catalysts are housed contains relatively large amounts of nickel. For the nickel content alone, it may become economical to remove the converters from junked automobiles and to recycle that material separately. In Japan, a law mandates this recycling to conserve nickel. Even if such removal and separate recycling become widespread in the United States, it is doubtful that recovery of the catalyst components would be economical, because of the very low concentrations of the platinum-group metals.

It is hoped that the recently discovered palladium–platinum ore in Montana can help to relieve our dependence on nondomestic sources of these materials. Because the catalytic activity is a rather insensitive function of the platinum : palladium ratio, it is likely that the catalysts can be made richer in palladium without sacrificing much in the way of activity. It has been suggested[136] that, for some gas scrubbing applications, palladium catalysts are better than those containing platinum.

EMISSION FROM STATIONARY SOURCES

Smelters

Because the platinum metals are so valuable, great care is taken to avoid losing significant amounts of them in refining processes. Thus, economic factors automatically tend to minimize health effects that might occur if compounds of these metals were to become distributed into the environment. Osmium appears to be the only platinum-group metal of which appreciable quantities are lost. In the refining of copper sulfide ores, osmium is converted to the volatile tetroxide.[65] It has been estimated[419] that 1,000–3,000 troy oz (31–93 kg) of osmium are lost in this process each year, but this relatively small amount does not appear to pose significant environmental hazards at present. It is recommended, however, that plants, animals, and humans living around copper smelters be examined for possible ill effects that might be attributed to accumulation of osmium in tissue.

Catalytic Uses

The only important stationary use of catalysts in which appreciable amounts of noble metals are lost in the reaction product is ammonia oxidation over a platinum–rhodium (90 : 10) gauze to make nitric acid. It is estimated that the amount of catalyst lost is usually in the range of 0.15–0.35 g/ton of nitrogen oxidized, although it may run as high as 0.5–1.8 g/ton in plants operating at higher pressures (e.g., up to 9 atm). Assuming an annual production rate of 40 million tons of nitric acid, one calculates that about 1.3×10^5 troy oz of platinum would be consumed in this process each year. Not all this material is allowed to escape into the atmosphere. Various filters and "gettering" devices now introduced into most of the plants allow recovery of perhaps half the lost metal. Of that which escapes, most is probably trapped in the absorption towers and ends up in the nitric acid. Through use of the nitric acid in fertilizers, in pickling baths, etc., the lost noble metals in the broadest sense are being "dumped" into the environment.

Platinum-metal catalysts have long been used in paint-drying ovens, wire-enameling ovens, and self-cleaning cookers.[2,185,289] In the earliest types, platinum metals were electrodeposited onto base-metal supports. These devices were designed to last 10 years with only semiannual washings required, if they were not chemically poisoned. Emission of the noble metals is negligible and poses no environmental hazard.

Flameless catalytic space-heaters have been labeled[80] an indirect threat to life, because of the dangerous concentrations of carbon monoxide that can be emitted from them. When they are used in unventilated spaces, the high carbon dioxide concentration and decreased oxygen concentration can combine with what would normally be an innocuous carbon monoxide concentration to produce harmful and even fatal results. Care must be exercised in using these devices. It does not appear that any catalytic material is emitted from them.

In addition to mining areas and noble-metal purification plants, the only other places where noble metals have been observed to reach physiologically significant concentrations are catalyst manufacturing plants. Although detailed information about concentrations in the air, soil, and employee tissues is not available, some statistics accumulated during the last 23 years of operation from a single such plant in Louisiana are given in the Appendix. Allergic reactions to platinum salts have been noted in a total of 15 people in the plant that now employs 250 people. Comparable allergic reactions would probably be observed in other plants that synthesize noble-metal catalysts.

9

Summary

Some new and potentially extensive uses of the platinum-group metals (platinum, palladium, rhodium, ruthenium, iridium, and osmium) have increased the possibility of environmental contamination with compounds containing these metals. Two of the new uses are in catalytic converters to control emission of pollutants in automobile exhaust and in drugs to arrest the spread of some types of cancer. The purposes of this report are to summarize what is known about the sources, properties, and uses of these metals and their compounds and the methods for analyzing them; to describe their toxicology; and to assess the environmental impact and possible health effects of their new uses.

Most of the platinum-group metals are produced from ores of ultrabasic rock formations mined in South Africa, Canada, and the U.S.S.R. Essentially none are currently produced from ores mined in the United States. Average concentrations of the six metals in the earth's crust are estimated to range from 0.001 to 0.01 ppm (1 ppm = 10^{-4} wt%), although they may be greater than 10 ppm in some commercially minable ore. Platinum and palladium are the most abundant of the six metals (>85%), and the ratio of the two varies from 0.3:1 to 2.5:1 in the various ores. Of approximately 4×10^6 troy oz of the metals produced annually, about 1.6×10^6 troy oz (40%) are consumed in the United States. The reserves outside the United States are estimated to be about 400×10^6 troy oz, enough to last about 100 years at current consumption rates.

Within the last 2 years, an extensive deposit of relatively concentrated ore (noble metals at about 20 ppm) rich in palladium (platinum:palladium ratio, about 0.3:1) has been discovered in Montana. The owners estimate that the deposit contains up to 500×10^6 troy oz of platinum-group metals, which is approximately equal to all the other known world reserves combined. This is the first potentially minable ore to be discovered in the United States, and the extent of its commercial possibilities is being studied.

After it is mined, the ore is treated in two distinct steps. The first stage (carried out near the mines) involves extraction of a concentrate of the precious metals from a large body of ore; the second stage (usually carried out in refineries at other locations) involves separation and purification of the individual metals. Two of these refineries are on the East Coast of the United States. Because of its intrinsic value, a large fraction of the used metal is recycled, or "toll refined," for further use.

The most common analytic techniques used to determine trace quantities of platinum-group metals are emission spectroscopy (preferred in Europe) and atomic absorption (preferred in the United States and Japan). The latter is more sensitive (platinum detection limit, about 0.1 ppm vs. about 10 ppm), but the former is somewhat less susceptible to physical and chemical interferences. Neutron-activation analysis can detect as little as 10^{-9} g of platinum and 10^{-11} g of iridium, which correspond to 0.001 and 0.00001 ppm, respectively, in a 1-g sample. Sample preparation can have a significant effect on the analysis, because the platinum-group metals are often not distributed homogeneously throughout a sample.

In a variety of ways, the chemical, petroleum, and electric industries account for about 80% of the platinum-group metals used in the United States. These metals are among the most versatile catalysts known. Before the introduction of catalytic converters on cars in 1975, the largest use of platinum was as a catalyst for the reforming of naphtha to make high-octane gasoline. Platinum, palladium, and rhodium are used as catalysts for oxidizing exhaust-gas pollutants, synthesizing nitric acid through ammonia oxidation, and hydrogenating a wide range of unsaturated hydrocarbons. Osmium and ruthenium are catalysts for stereospecific hydroxylation of olefins, and ruthenium has high catalytic activity for formation of polymethylenes from carbon monoxide and hydrogen at high pressure. Portable space-heaters using platinum catalysts are also becoming popular.

Because of its resistance to oxidation, palladium (sometimes alloyed with other platinum-group metals) is used extensively in the electric industry for contacts in relays and switchgear, in resistors and capacitors, as electrochemical and spark electrodes, and in fuel cells. Platinum (also frequently alloyed with other metals) has many high-temperature applications, such as in precision resistance thermometers, thermocouples, strain gauges, and laboratory ware. Palladium has the remarkable ability to dissolve large quantities of hydrogen, a property that makes it useful in devices to purify hydrogen. Because of their nontarnishing reflective properties, the platinum-group metals are used as reflector surfaces and in jewelry.

Other properties that make the platinum-group metals useful are their resistance to oxidation, ductility, high melting points, good electric conductivity, and high density. They form a variety of alloys with each other and with other metals that can greatly affect their hardness.

In the metallic state, the platinum-group metals are normally unresponsive to direct chemical attack by oxygen, halogens, acids, etc., under mild conditions; however, at high temperatures (about 1000° C), volatile oxides and halides can be formed. The metals are more susceptible to attack by alkaline fusion, especially in the presence of oxidizing agents. The state of subdivision of the metal particles can have a profound effect on their reactivity, the smaller particles normally being the more reactive.

All the platinum-group metals form water-soluble organometallic coordination compounds. Some of these complexes are now being used as highly selective homogeneous catalysts in commercial applications (e.g., rhodium halides for converting methanol and carbon monoxide into acetic acid with selectivity greater than 99%). Others, such as the neutral square-planar *cis*-dichlorodiammineplatinum(II), have physiologic activity and are being developed as anticancer drugs. Cluster compounds containing two or more of the platinum-group-metal atoms per molecule bound together with bridging species, such as carbonyl groups, can also be synthesized. These compounds permit variation of the metal–metal bond distance, a factor that may have a significant influence on catalytic behavior.

In their metallic states, the platinum-group metals have extremely low toxicities. Some alloys containing these metals are actually used as dentifrices and prostheses. However, many of the soluble compounds of these metals (e.g., chlorides, aminochlorides, and volatile oxides) are highly toxic and can elicit physiologic responses at concentrations as low as 10^{-9} g/ml. Generally, the chlorides are more toxic than the oxides. Within a given class of compounds, the toxicity appears to follow the water solubility, although lipid solubility is also necessary to effect transport into the cells. The route of administration affects a compound's toxicity and decreases in this order: intravenous > intratracheal ~ intraperitoneal > inhalation >> oral. For example, the LD_{50} for $PdCl_2$ in rats and rabbits is about 5 mg/kg of body weight when administered intravenously, but 200 mg/kg when given orally. Neither PtO nor $PtCl_2$ causes ocular irritation, although a few milligrams of $PdCl_2$ will cause corrosive lesions of the conjunctiva and severe inflammation of the cornea.

Several ionic derivatives of the platinum-group metals have the capacity to become attached selectively to specific chemical sites in proteins. For example, $PtCl_4^{2-}$ will attack such functional groups as disulfide bonds, terminal $-NH_2$ groups, and methionines. Isomorphic replacement with the heavy metal allows the structure of proteins to be determined more readily by x-ray crystallography. $PdCl_2$ selectively combines with and renders inactive the enzymes trypsin and chymotrypsin, although it does not inhibit catalase, lysozyme, peroxidase, and ribonuclease. Platinum complexes also interact with functional groups (e.g., $-NH_2$, $-CO_2^-$, and $-SCH_3$) on amino acids and with many viruses. They are highly effective in deactivating bacteria, although neuromuscular toxicity prevents their general clinical use in combating bacterial infections.

Of particular significance is the ability of platinum complexes to bind strongly with receptor sites in nucleic acids, such as DNA. It is thought to be this ability that makes the complexes effective as anticancer drugs. *cis*-Dichlorodiammineplatinum(II), as well as other similar *cis*-dichloro complexes, is highly effective in causing regression of tumors in animals. Now in Phase II clinical trials, these drugs also have shown effectiveness in treating cervical, testicular, ovarian, prostatic, bladder, and head and neck cancer in humans. When used by themselves, the drugs have very severe side effects on kidneys; however, these effects can be lessened by proper selection of the ligands in the platinum complex and by pharmacologic treatment of the patient. *cis*-Dichlorodiammineplatinum(II) is a useful drug in the treatment of some types of neoplasm; it appears most effective when used in combination with other antineoplastic agents. It is anticipated that additional analogues with higher therapeutic ratios will be developed.

Compounds of platinum-group metals are quickly eliminated from the body (mainly in the urine and feces), except if administered intravenously, in which case elimination is much slower. The metals retained are apparently concentrated in the kidneys, liver, and spleen.

The effects of inhalation have become of interest, because that is the most likely mode of exposure of the general population if significant

Summary

quantities of the metals are emitted from exhaust systems of cars equipped with catalytic converters. Present exposure is limited mainly to employees working in platinum-metal refineries who inhale dust particles of salts or acids of these metals. Such people often become "sensitized" and develop asthmatic or dermatologic allergies (sometimes called "platinosis") that disappear when the exposure is discontinued. These reactions are thought to be caused by histamine release triggered by the compounds. Only ionic water-soluble platinum salts cause these reactions.

A very sensitive and reliable skin-prick test has been devised to separate people into two distinct classes: atopic (sensitive to common environmental allergens—about 30% of the population) and nonatopic (the remaining 70%, who are less susceptible to allergens). After prolonged exposure by inhalation of dusts containing the soluble platinum salts, a significant fraction (over 50%) of people from both classifications can become "sensitized," i.e., show clinical symptoms of allergic reactions; the atopic group become sensitized more quickly than the nonatopic. As with the toxic behavior, the allergenicity of the ionic platinum complexes is apparently determined by the number of chlorine atoms in the molecules; ammonium chloroplatinate, $(NH_4)_2PtCl_6$, and chloroplatinic acid, H_2PtCl_6, are the most allergenic of the compounds tested. Exactly what concentrations are required for sensitization and for eliciting clinical responses in sensitized people is not known; however, it is clear that much lower concentrations can cause responses in sensitized people than are required to cause the initial sensitization. Once sensitized, a person apparently retains that tendency for the rest of his life.

Environmental exposure to soluble salts of platinum-group metals is currently confined primarily to mining areas, platinum-metal refineries, and catalyst synthesis plants. Measurements of soil near freeways and tissue samples from autopsies of unexposed people show noble metals at less than 1 ppb (1 ppb = 10^{-7} wt%). Air samples taken near freeways in California typically contain platinum or palladium at less than 10^{-7} $\mu g/m^3$ (the detection limit of the analytic techniques used). Autopsy samples of people occupationally exposed to the platinum-group metals through employment in mining areas also show negligible concentrations of the metals in body tissue. However, several urine samples from refinery workers have shown measurable concentrations of platinum and palladium, although concentrations in the blood of such workers were below the detection limits of the analytic equipment. Apparently, the inhaled salts are rapidly excreted, primarily in the urine, and are not accumulated in the body tissue, although measurements of tissue samples from autopsies of refinery workers need to be made.

The extremely low concentrations of the platinum-group metals make it difficult to obtain reliable baseline data. Sample contamination must be rigorously avoided. The difficulties are readily illustrated by the inconsistencies in concentrations reported by the various laboratories, which can vary by much more than an order of magnitude. Standardized procedures need to be developed, with frequent cross-checks on the same sample by various laboratories involved in making the measurements.

Catalysts containing small amounts of platinum, palladium, and rhodium to remove pollutants from automobile exhaust were introduced in most 1975 cars sold in the United States; these devices were necessary to meet the stringent limits on maximal emission of carbon monoxide, hydrocarbon, and NO_x mandated by the Federal Clean Air Act of 1970. In the converters, platinum and palladium (in a ratio of about 2.5:1, which is typical of the ratio obtained in the South African mines) at about 0.06 oz/car are deposited as small crystallites on the surface of spherical pellets (or along the channel walls of ceramic monolithic supports) made of a refractory oxide, such as alumina.

Carbon monoxide and hydrocarbon (the only pollutants now being catalytically controlled) require an oxidizing atmosphere, which is achieved by running the engine "fuel-lean" or by adding excess air to the exhaust stream via a fanbelt-driven air pump. Catalytic removal of NO_x, the most difficult pollutant to control, requires a reducing atmosphere (accomplished by running the engine "fuel-rich"). Normally, the conversion of all three pollutants would require at least *two* separate catalyst beds, each having a unique chemical composition and involving a different type of atmosphere. However, if the fuel composition can be maintained within about ±0.2 air:fuel ratio units of the stoichiometric value necessary for complete combustion, all *three* pollutants can be controlled in a *single* three-way catalyst bed. It is likely that, as the NO_x standards become increasingly stringent, this will be the approach taken to comply.

The use of oxidizing catalytic converters has contributed greatly to decreasing the emission of carbon monoxide and hydrocarbon, including such species as benzo[a]pyrene, which are known to be carcinogenic. That control of such emission improves air quality has been dramatically illustrated by a comparison of the effects of diluted exhaust from catalyst and noncatalyst cars on the health of several species of mammals. Survival rates of infant rats and the general health

Summary

of animals exposed to noncatalyst exhaust were poor, whereas those exposed to catalyst exhaust behaved the same as control groups subjected to normal air. Moreover, the catalytic converters permit increased fuel economy, because they effectively decouple emission control from engine performance and allow the latter to be optimized.

On the negative side, oxidizing catalytic converters have a tendency to convert part of the fuel sulfur into sulfuric acid aerosol mists with particles in the respirable size range. Even though the vehicles are responsible for only 1% of the total sulfur oxides released by human activity in the United States, mathematical models developed by the EPA have predicted that, under "worst-case" conditions (heavy vehicular traffic and adverse meteorologic conditions), locally high concentrations (up to 600 $\mu g/m^3$) might occur in the vicinity of freeways. The threshold for irritability in sensitive people (on the basis of 24-h exposure) is thought to be around 10 $\mu g/m^3$. These mathematical models, however, are not very reliable and tend to overestimate the problem. Wind, even at velocities as low as 0.2 mph, can dramatically decrease the maximal concentration, and chemical reactions to form sulfates (which are less toxic than sulfuric acid) lower the steady-state acid concentration further. Under any realistic conditions, it is doubtful that the "worst-case" concentrations of sulfuric acid will exceed about 50 $\mu g/m^3$.

Recently, experimental tests with 352 cars operated under simulated freeway conditions gave observed sulfuric acid–sulfate concentrations much lower (by at least a factor of 7) than would have been predicted by some of the early EPA models. Particularly significant was the observed vertical displacement of the acid–sulfate plume, which resulted in concentrations lower than the models predicted near the ground.

Although concentrations of sulfuric acid and sulfates around freeways are indeed increased because of the presence of catalytic converters on cars, it appears that the hazard is not nearly so great as originally feared.

It is generally accepted that minute quantities of noble metals are emitted from catalytic converters, perhaps 1–3 μg/mile (about 0.6–1.9 μg/km). There is no evidence that any of this emitted material is in the form of physiologically active soluble platinum-metal salts. Moreover, even if all the emitted material remained suspended in the air and behaved like other aerosol particles, the maximal possible steady-state concentration (according to the EPA "worst-case" model) would not exceed 0.06 $\mu g/m^3$; the OSHA standard for exposure to soluble platinum salts is 30 times larger than that. However, if all the emitted material were localized in the topsoil near the freeways, it would take over 1,000 years for it to reach the concentrations observed in the ore in the South African mines. Although platinum(IV) compounds can be methylated by microorganisms, there is no evidence to suggest that platinum metal (the probable chemical form of the emitted material) can be methylated in such a way. It is thus concluded that emission of platinum-group metals from catalytic converters in cars will not pose a significant health problem. However, it would be appropriate for the concentration in soil near busy freeways to be monitored during the next several years.

Overheating of catalytic converters due to engine malfunction sometimes occurs and has led to a small number of fires both inside vehicles and in grass beneath the cars. However, this problem is not peculiar to cars with catalytic converters; many late-model cars without catalytic converters also have exhaust systems that are hot enough to cause grass fires. This tendency makes it important to keep the engines of automobiles in good working condition and to exercise extreme care when operating cars in such areas as fields and national parks, where there is dry grass.

The use of platinum-group metals in catalytic converters has increased by nearly 50% the demand for these imported materials in the United States. Production from foreign mines has been expanded to meet this demand, but installation of such devices on cars in western Europe and Japan could place a burden on the traditional suppliers. It is hoped that the newly discovered deposits in the United States can be developed to meet some of this demand.

It is concluded that at present the only adverse health effects directly attributed to the platinum-group metals involve workers in refineries that purify the metals and plants that synthesize noble-metal catalysts. However, it is suggested that vigilance be maintained to determine whether problems arise as a result of emission of harmful compounds into the environment from catalytic converters or from anticancer drugs that contain these metals.

10

Conclusions

TOXICITY CLASSIFICATION

Some of the water-soluble complex salts (e.g., chlorides and aminochlorides) and volatile oxides (e.g., OsO_4, RuO_4, and IrO_2) of the platinum-group metals have physiologic activity; chlorides are more toxic than oxides. Within a given class of compounds, the toxicity apparently follows the water solubility to some degree. Although water solubility is necessary for toxicity, it is not by itself a sufficient condition; lipid solubility is also necessary, to effect transport into the cell. Nonvolatile oxides (e.g., PtO and PtO_2) have very little toxicity or allergenicity. None of the platinum-group metals is toxic or allergenic in the metallic state.

ROUTE OF ADMINISTRATION

For a given compound, toxicity varies with route of administration. Usually (e.g., for $PdCl_2$), the order of toxicity is: intravenous > intratracheal ~ intraperitoneal > inhalation >> ingestion; in some cases (e.g., $RhCl_3$), intraperitoneal administration produces more toxicity than does intravenous. Absorption into the bloodstream may be the most important factor in determining toxicity.

MODE OF CIRCULATION

Almost all (over 99%) of toxic platinum compounds circulates as protein-bound material in the serum; almost none is absorbed by the tissues. Although large amounts of the compounds enter the placenta, only small amounts enter the fetus in rats. Furthermore, almost none enters the brain.

ELIMINATION

Elimination of metal salts administered orally in trace or pharmacologic quantities is rapid (>99% in 3 days; more in the feces than in the urine). When they are administered intravenously, elimination is much slower; only about half the material is excreted in 14 days. The retained material is concentrated mainly in the liver, kidneys, and spleen.

CAUSES OF ALLERGENICITY

Several soluble platinum compounds can cause allergic responses in sensitized people. These allergic reactions are apparently caused by histamine release triggered by the platinum compounds. Antihistamines can protect animals (e.g., guinea pigs) and humans from anaphylactic shock resulting from sodium chloroplatinate in either the +2 or the +4 oxidation state. However, there is apparently no relationship between histamine release and acute toxicity.

SKIN-PRICK TESTS

Skin-prick tests with dilute concentrations of soluble platinum complexes appear to provide reproducible, reliable, and highly sensitive biologic monitors of allergenicity. After sensitization through previous exposure, administration of as little as 3×10^{-15} g of the allergenic compound will produce a wheal in highly susceptible people. There are no suitable animal models by which to measure human response to these compounds.

CANCER CHEMOTHERAPY

Platinum-group-metal complexes—e.g., *cis*-dichlorodiammineplatinum(II), 1,2-dinitratodiamminecyclohexaneplatinum(II), and rhodium(II) carboxylates—have remarkably high therapeutic indexes (some as high as 500 in selected animals) as chemotherapeutic agents for arresting some types of cancer. The biggest problem with these compounds in humans has been their harmful effect on the kidneys. However, this difficulty has been substantially decreased by selection of appropriate ligands in the complex and by applying the complexes in combination with other drugs. With these successes, some of the most promising compounds are now being tested in Phase II clinical trials, and it is anticipated that they will become widely used as anticancer agents in the near future.

REFINERIES

The only places where platinum-group metals reach physiologically noticeable concentrations are in refineries that purify the metals, mines, and plants that synthesize catalysts containing the metals. A significant fraction (around half) of the people exposed in these refineries to some water-soluble salts of platinum develop allergic reactions, with symptoms similar to those of hay fever, and dermatologic problems. Although the manifestations are not usually so severe that they require corrective action, in a few cases the exposed people must be moved to job locations where such exposure does not occur. When this is done, the allergic manifestations disappear.

CATALYTIC CONVERTERS

The newest and most extensive use of the platinum-group metals is in catalysts for purifying exhaust from automobiles. Minute quantities of platinum and palladium (about 1–3 μg/mile, or 0.6–1.9 μg/km) are emitted from the exhaust systems of automobiles equipped with catalytic converters; much of this material may accumulate alongside roadways. However, this material is in a chemical form that is physiologically innocuous (no detectable soluble salts), and it is concluded that such emission poses no threat to the environment. Because there is no evidence that platinum metal can be methylated by microorganisms and solubilized in the same way that mercury is methylated, this deposited material should not have an adverse effect on the environment.

SULFURIC ACID

Health effects due to emission of sulfuric acid from automobiles equipped with catalytic converters that contain platinum-group metals will probably not be very serious. Experiments conducted under simulated freeway-driving conditions have demonstrated that the concentrations of sulfuric acid and sulfate are much lower (perhaps by more than an order of magnitude) than those originally predicted on the basis of mathematical modeling. It is concluded that available information about the effects of turbulence, dispersion, thermal drafts, atmospheric reactions, surface adsorption, and wind velocity is insufficient to permit valid predictions from such models.

PLATINUM RECYCLING

Recycling of platinum-group metals used as catalysts in the petroleum and chemical industries is substantial. Recovery of metal returned to the refiner usually exceeds 99%. It is doubtful, however, that there will ever be a significant reclamation of used metals from automobile emission control catalysts, because of the relatively low value of the metals (<$10 in each converter), the low concentrations (<0.1 wt%), the high cost of dissolving the refractory oxide support, and the logistics of returning the material to the central refinery.

MATERIAL AVAILABILITY

Material availability may become a problem, because almost none of the platinum-group metals is produced from domestic sources. Most of it comes to the United States from mines in South Africa, the U.S.S.R., and Canada. The recent introduction of catalytic converters in automobiles has almost doubled our use of these materials. However, the recent discovery of a new deposit of palladium-rich ore in Montana could help to relieve the situation. It is estimated that the worldwide proven reserves (excluding the new U.S. discovery) should be sufficient to last about 100 years at current rates of consumption, although the consumption will probably increase substantially as other countries (e.g., Japan) begin to require catalytic converters for control of emission from automobiles.

11

Recommendations

TESTING OF COMPOUNDS

Newly synthesized compounds containing platinum-group metals should be routinely tested for acute and chronic toxicity, for allergenicity, and for therapeutic potential as possible agents for cancer chemotherapy. In addition, new compounds (as well as several known compounds, particularly those which could enter the biosphere) should be subjected to long-term physiologic testing for carcinogenicity and teratogenicity. Bioassays should be established to determine whether accumulation of these compounds occurs.

BASELINE TISSUE DATA

As platinum-group compounds become commonly used as cancer chemotherapy agents, the materials will increase in the tissues of treated patients. For this reason, it will become important to monitor their concentrations continuously and to seek concentration trends in the overall population. Current baseline tissue data are contradictory, and observed concentrations of these metals vary over two orders of magnitude. Reliable, standardized methods for ascertaining tissue concentrations need to be developed, and the same material should be measured by different groups, to ensure consistency. At the very low concentrations in question, contamination takes on critical importance. Baseline data need to be collected on workers in platinum-metal refineries and in catalyst-synthesis plants.

ALLERGENICITY VS. ATMOSPHERIC CONCENTRATION

The inhalation tests developed to examine sensitivity to airborne particulate matter, although reasonably reproducible, cannot be related easily to concentrations of the same materials in the air. Because inhalation of particulate matter is an important route of exposure of people in refineries (and possibly of the general population, if significant quantities of metal are emitted from catalytic converters), it is important to determine the effects of concentration, chemical composition, and particle size. More refined tests to examine these effects need to be developed. This information is critical in establishing useful concentration standards.

LEADED GASOLINE AND CATALYTIC CONVERTERS

It has been concluded that the amounts of platinum and palladium normally emitted in exhaust from cars equipped with catalytic converters are very small. However, the measurements were made with fuels that contained no lead. In practice, leaded gasoline is occasionally introduced (whether inadvertently or deliberately) into these vehicles, and the consequences of such introduction have not been fully investigated. Although the poisoning of the activity by lead has been well documented, the effects of the halide scavengers (ethylenedichloride and ethylenedibromide) on the volatility of the noble metals are not known. It is possible that the halides react with the metals to form volatile species that can lead to increased removal of the metals from the converters. This possibility needs to be tested.

ALLERGENICITY AND CATALYTIC CONVERTERS

People who have been "sensitized"—i.e., who exhibit allergic responses when exposed to very small quantities of some soluble platinum compounds—are extremely sensitive biologic monitors of trace amounts of these materials. It is recommended that carefully conceived tests be made with such hypersensitive people, to seek evidence of emission of these compounds from cars equipped with catalytic converters. Although almost all of the small amount of platinum-group metals is emitted in an insoluble form, these tests might provide the most sensitive diagnostic tool for measuring emission of soluble material. This may become important to the relatively few people who are hypersensitive.

NEW SOURCES

The claim of an extensive deposit of palladium-rich ore in Montana needs to be confirmed, and its market potential evaluated. The search for additional deposits of ore in the United States needs to be encouraged.

Appendix: Catalytic Reforming in Refining of Petroleum

GENERAL OVERVIEW

The use of platinum catalysts in petroleum-refining has been discussed in National Materials Advisory Board (NMAB) Report 297 (March 1973), *Substitute Catalysts for Platinum in Automobile Emission Control Devices and Petroleum Refining.*[298] Considerable changes have since taken place in the refining operation, particularly with the introduction of the mandatory use of unleaded gasoline in 1974. The changes in gasoline production can be seen in Figure A-1, which shows the production of total gasoline and premium gasoline from 1968 to 1974. It is interesting to note that the discontinuity in gasoline production in 1971 is coincident with the impact of the initial noncatalytic attempts at pollution control—lower compression ratios, higher air : fuel ratios (engines run with leaner carburetion, or with excess air), and greater spark-advance retard. As the demand for premium gasoline diminished, a portion of the capacity for making the premium gasoline went into production of unleaded gasoline, when the need for unleaded gasoline became effective in 1974. Thus, the total refinery balance, with respect to a pool octane, has not been greatly affected. With the introduction of catalytic converters on 1975-model cars, a fuel-economy benefit has been realized, and there are now economic pressures to improve this further. Although the direction toward lighter vehicles is obvious, the sacrifice in car size and comfort may be more

Appendix: Catalytic Reforming in Refining of Petroleum

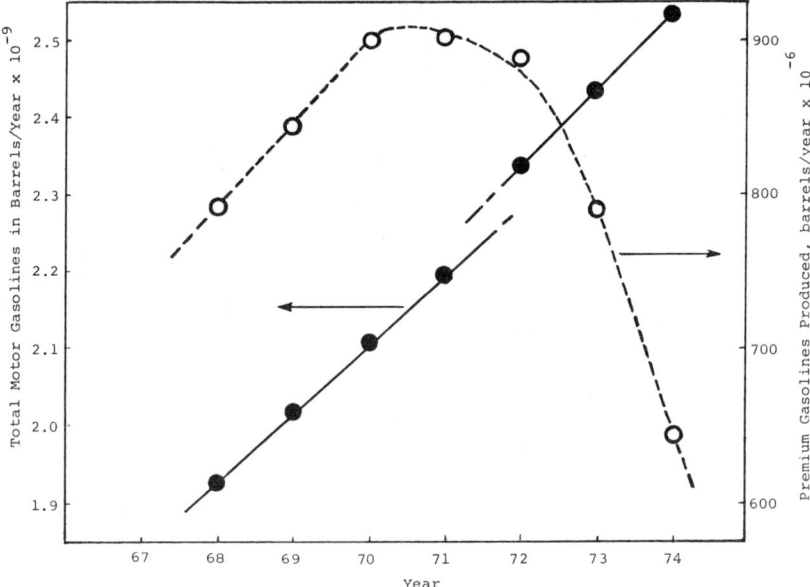

FIGURE A-1 Total production of gasoline and premium gasoline in the United States, 1968–1974. Data from Universal Oil Products Company.

difficult to sell in a competitive automotive market, so other methods of increasing economy must be examined. The most obvious one is a higher compression ratio, which provides for a higher cycle efficiency and some reduction in engine weight; both factors are beneficial with respect to fuel economy (see Figure A-2).

Higher compression ratios will lead to a greater emission of nitrogen oxides; however, within the period involved, it is quite likely that better carburetion devices for controlling the air : fuel ratio will be developed for use in conjunction with the catalytic converters. Thus, a truly three-way control system for carbon monoxide, hydrocarbon, and NO_x will have a positive impact on both fuel economy and environmental quality.

What will be the impact on the refining industry? Within the next 5 years, the industry will probably be heading in the direction of producing a high-octane unleaded fuel as a major refinery product. The gain in fuel economy will more than offset the crude-requirement penalty associated with making an unleaded, high-octane pool gasoline. From Figure A-3, it will be observed that, for a given number of miles driven,

Appendix: Catalytic Reforming in Refining of Petroleum

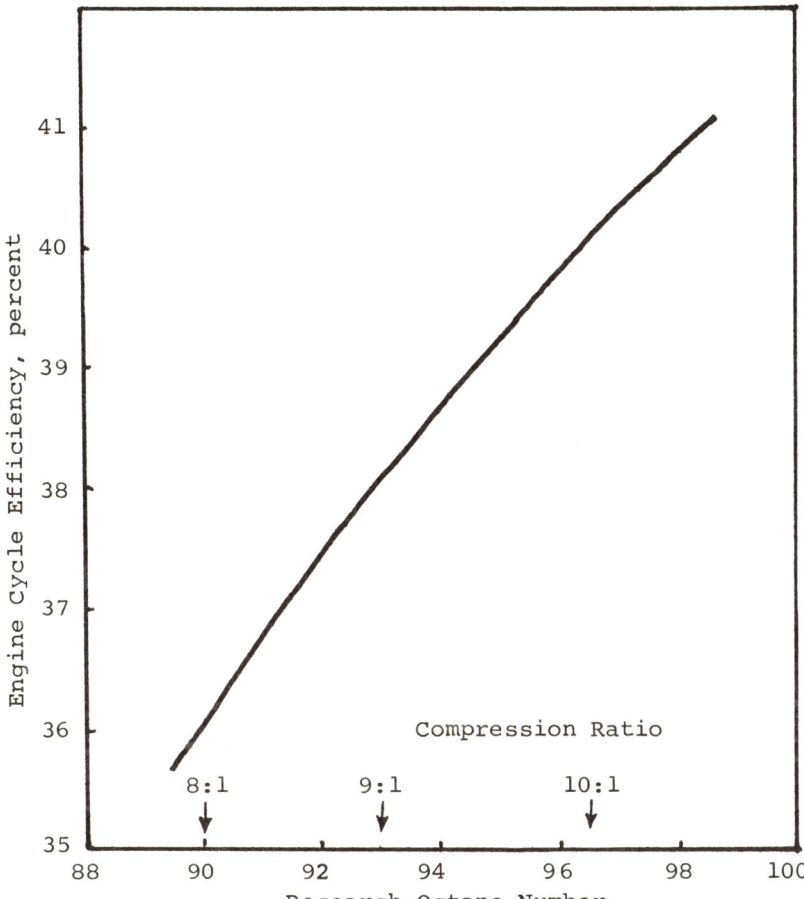

FIGURE A-2 Relationship between research octane number, engine cycle efficiency, and compression ratio. Data from Universal Oil Products Company.

the crude requirement is at a minimum when the pool octane number is about 96.5. It should be noted that the crude-requirement change depicted in Figure A-4 merely indicates an additional crude requirement for a given volume of gasoline; in reality, because some of that additional crude is converted into other products (such as methane and liquid petroleum gas), the byproducts are salable or useful for internal refinery energy needs. The net additional refinery energy need to finance the higher pool octane number is around 1%.

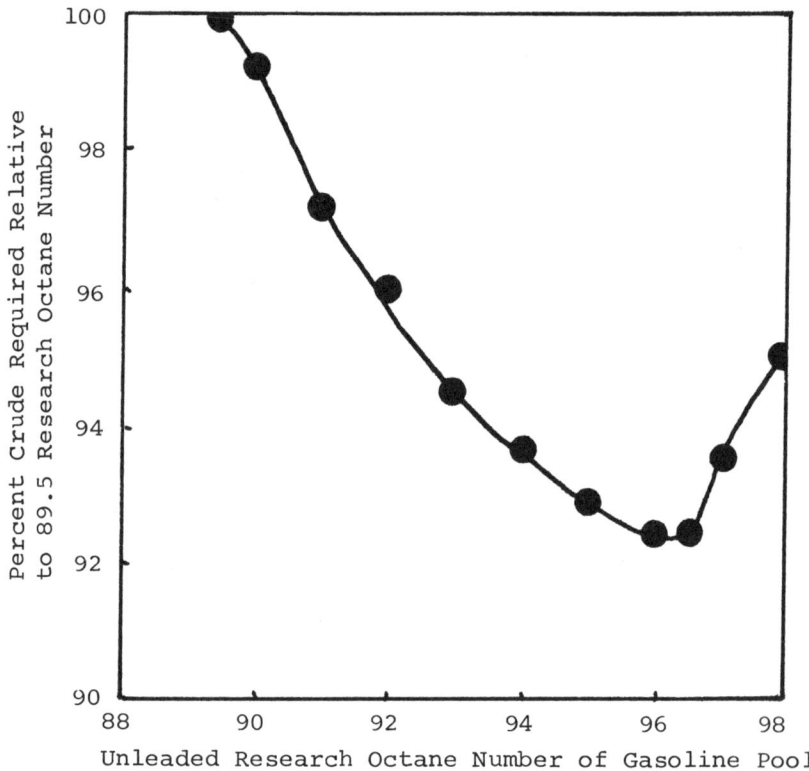

FIGURE A-3 Relative crude-oil requirement to drive a given car a given distance under same driving conditions with unleaded gasolines of research octanes above 89.5. Data from Universal Oil Products Company.

In the petroleum-refining industry, the major use of platinum catalysts is in catalytic reforming. The last 5 years have witnessed a switch to bimetallic catalysts for reforming. A number of combinations are in commercial use; of these, platinum–rhenium is one that is widely used. The use of bimetallic catalysts represents a great improvement in platinum-based catalytic reforming systems, from the standpoint of efficient and stable operation. In addition, there has been a reduction of around 20% in the concentration of the platinum component in bimetallic catalysts. It is estimated that the platinum used in catalytic reforming is now over 1,100,000 troy oz (34,200 kg) and increases by about 200,000 troy oz (6,200 kg) each year.

With the increased use of bimetallic catalysts and the more sophisticated performance requirements, the chances of substitution for these

Appendix: Catalytic Reforming in Refining of Petroleum 185

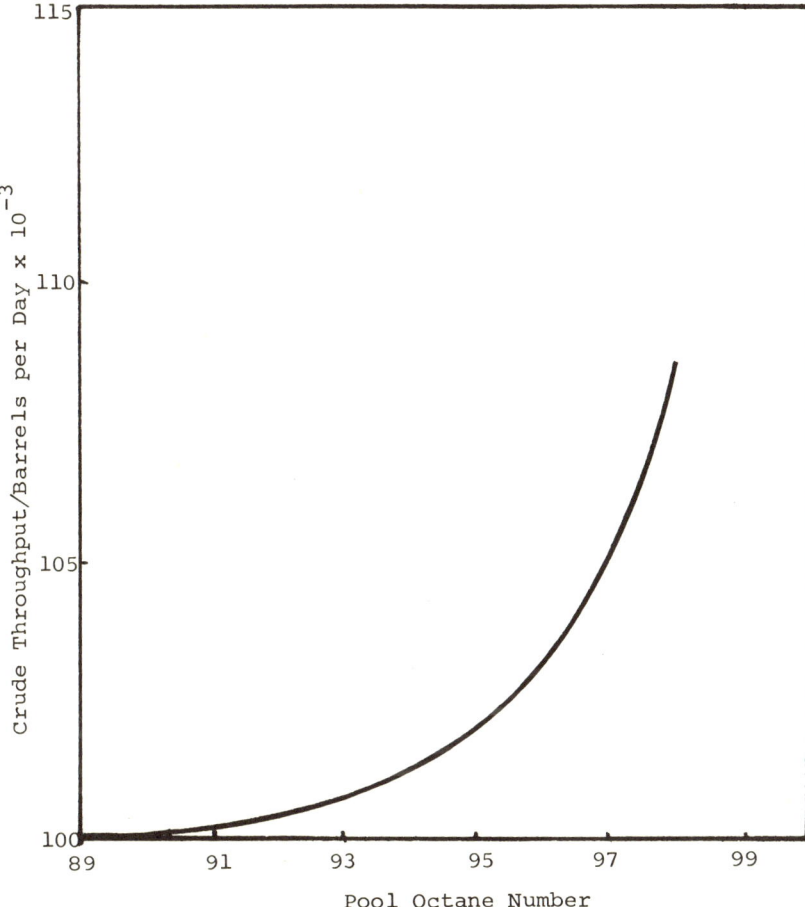

FIGURE A-4 Estimated crude-oil requirements for various pool octane gasolines. Data based on constant product distribution, East Coast location, 100,000-barrel/day capacity, and typical refinery. Data from Universal Oil Products Company.

catalyst systems in catalytic reforming appear to be even more remote than they were at the time of the issuance of the NMAB report on substitute catalysts.

Some comments on precious-metal recovery and allergic reactions to platinum salts are appropriate.

Because the platinum-group metals are so valuable, great care is exercised in monitoring all streams entering and leaving the catalyst manufacturing sites, to ensure that no significant amounts of materials are lost. Yearly overall material balances of about 99.75% are attainable.

Regarding allergic reactions to platinum salts, an example can be drawn from records about employees maintained since the commissioning of the platinum-catalyst manufacturing plant by the UOP Company in Shreveport, Louisiana, in 1952. The work force was 100 persons at that time and has grown to 250 persons. During 23 years of operation, a total of 15 persons developed allergies to platinum salts, with various degrees of sensitivity. In a number of instances, removal to a different location or job in the same plant area solved the problem. Those with the greatest sensitivity have moved to jobs outside the company. There have been no reports of allergic reactions when the catalyst in its final form was handled at the refinery site.

The NMAB report (p. 2 of "Summary of Conclusions and Recommendations") stated that "the use of platinum catalysts in other petroleum processes also will increase, but until 1980 the amount of platinum required for these processes will be less than 25% of that needed for reforming." This statement has been reexamined in the light of the present trend toward clear gasoline production. The other processes that use platinum and palladium are hydrocracking, isomerization, and hydrogenation (benzene to cyclohexane). The balance of the various refinery processes will depend to a great extent on the characteristics of the crude-oil supply and the changing needs for various products, as well as on improvements in refining. Nevertheless, the emerging patterns are such that isomerization will be playing a more important role in gasoline production, in view of the remaining need to balance out the relationship between boiling point and octane number. Hydrocracking is playing an important role in the production of clean middle oils and may assume greater interest for gasoline production; however, the advances in catalytic cracking have a bearing on the selection of the process for gasoline manufacture. With respect to the use of noble metals, these appear to be much more entrenched in isomerization than in hydrocracking, where the trend has been away from noble metals. Thus, it remains safe to suggest that, although the process split has been changing, noble-metal use in nonreforming processes will remain below 25% of reforming use for a number of years.

CATALYTIC REFORMING

In the petroleum industry, the word "reforming" is used to designate a process by which the molecular structures of naphthas are changed, or "reformed," with the intent of lessening their knocking tendency (or raising the octane number, which describes the ability of the fuel to burn smoothly under conditions of high pressure) in internal-combustion engines. The reforming process is also used to synthesize aromatics—particularly benzene, toluene, and C_8 aromatics—from selected naphtha fractions.

The antiknock quality of unleaded gasolines is related to the chemical structures of their constituent hydrocarbons. Paraffins, olefins, naphthenes, and aromatics are the four main hydrocarbon types of which gasolines are composed. Normal paraffins have the lowest octane numbers among the hydrocarbons, and the isomerized or branched paraffins have much higher octane ratings. It is well known that the octane number scale has been defined by ascribing a zero rating to n-heptane, which is particularly prone to knocking, and a rating of 100 to isooctane (2,2,4-trimethylpentane), one of the more highly branched octanes. Generally, monoolefins have higher octane numbers than corresponding paraffins. Naphthenes, or cycloparaffins, have very high octane numbers. The aromatics have exceptionally high octane numbers, generally over 100. Although the relationship between hydrocarbon structure and knock rating is highly involved, these broad generalizations indicate the structural changes that reforming processes are intended to accomplish, to raise the octane number of gasolines.

Native hydrocarbon types vary widely in the proportions in which they occur in petroleum from different fields; therefore, the octane ratings of "straight-run" gasolines vary. Most straight-run gasolines, obtainable by simple distillation from crude oil, contain only paraffins, naphthenes, and aromatics, and they have octane numbers of 50 or less. As examples, the hydrocarbon types and octane numbers of a typical domestic midcontinent naphtha (produced from the center of the United States) and of a Kuwaiti depentanized naphtha are shown in Table A-1. The octane number of straight-run naphthas, even with the addition of lead alkyls, is too low to permit their inclusion in commercial gasolines. Therefore, the chemical compositions of these straight-run naphthas need to be changed, or reformed, so that they can be used in modern internal-combustion engines without knocking.

THERMAL REFORMING

The first refining process used to change the composition of native naphthas to improve their octane rating was thermal reforming, introduced in 1930. In this process, which was conducted at just over 1000° F (538° C) and at pressures of 500–1,000 psig (3,450–6,900 kN/

Appendix: Catalytic Reforming in Refining of Petroleum

TABLE A-1 Compositions and Octane Numbers of Straight-Run Naphthas[a]

	Source	
	Midcontinent	Kuwait
Hydrocarbon composition, vol%:		
Paraffins	48	67
Naphthenes	42	22
Aromatics	10	11
Research octane number:		
Clear	47	39
With tetraethyl lead at 3 ml/gal	73	67

[a]Data from Sterba.[428]

m^2), olefins were produced from paraffins, high-molecular-weight paraffins were cracked to produce low-molecular-weight paraffins with higher octane numbers and olefins, and native aromatics were concentrated by the cracking of paraffins into much smaller gaseous fragments. Although other reactions were involved, those named resulted in an increase in the research octane number up to about 85, but the liquid product reformate yields would have been too low to be economical today. The creation of aromatics with very high octane numbers was insignificant in thermal reforming; they were concentrated by the destruction of paraffins to gaseous hydrocarbons.

The inefficiency resulting from this destruction of paraffins by the thermal process and the inability to synthesize aromatics constituted an incentive to develop a catalytic reforming process that could be more selective in promoting the desired reactions and in minimizing the unwanted reactions. The first successful catalytic process that arose from this effort made its commercial appearance in 1949 under the name of the "UOP Platforming process"; it used a catalyst containing a noble metal.

PROCESS DESCRIPTION

Principal reactions (to be described later) that characterize reforming processes that use noble-metal catalysts are:

- dehydrogenation of naphthenes to aromatics, to near completion, with very little ring rupture

Appendix: Catalytic Reforming in Refining of Petroleum 189

- isomerization of paraffins to more highly branched forms
- dehydrocyclization of paraffins to aromatics
- hydrocracking of paraffins to lower molecular weights, but with minimal production of light gaseous hydrocarbons

Because these reactions are conducted in an atmosphere of hydrogen under pressure, no olefins are produced; catalytic reformates are therefore much less susceptible to oxidative reactions and are more stable in storage than thermal reformates. The hydrogen environment in the catalytic reaction zone is created deliberately to minimize fouling and deactivation of the catalyst by the formation of carbon on its surface.

The superiority of catalytic over thermal reforming was demonstrated experimentally in a comparison[177] that showed a reformate yield of 85 vol% for the catalytic process and a yield of only 55 vol% for the thermal process, in the reforming of a midcontinent naphtha to a product with a research octane number of 85.

The superiority of catalytic over thermal reforming was demonstrated experimentally in a comparison[177] that showed a reformate yield of 85 vol% for the catalytic process and a yield of only 55 vol% for the thermal process, in the reforming of a midcontinent naphtha to a product with a research octane number of 85.

Although a variety of catalytic reforming processes are in use, the simplified flowsheet in Figure A-5 describes the Platforming process, which was the original of the family using noble-metal catalysts and is still the most extensively used process of this group. The flowsheet shows the reforming unit to be composed of five principal sections:

- reactors that contain the particulate catalyst in fixed beds
- heaters to bring the hydrotreated naphtha charge and hydrogen recycle gas to reaction temperature and to supply the heat of the reaction
- product cooling section and a gas–liquid separator
- hydrogen-gas recycling system
- fractionation to separate light hydrocarbons dissolved in the separator liquid

The particulate catalyst is contained as a fixed bed in three (sometimes more) separate adiabatic reactor vessels, with combined-feed preheating before the first and reheating between later reactions. Because of the rather large endothermic heat of the dehydrogenation reactions, there is a substantial drop in temperature of the flowing reacting stream, particularly in the first reactor, in which the rapid naphthene dehydrogenation reaction occurs. Therefore, the effluent

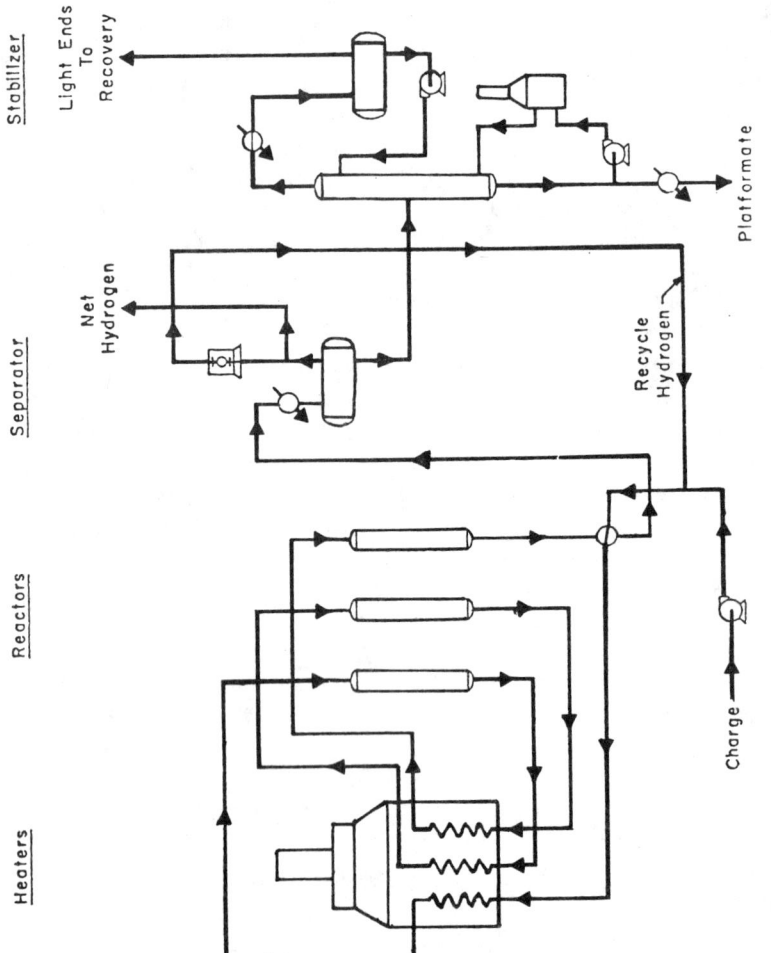

FIGURE A-5 Universal Oil Products Company conventional Platforming process.

Appendix: Catalytic Reforming in Refining of Petroleum 191

from the first and second reactors is reheated to the required inlet temperature for the subsequent reactor. Usually, the charge heater and the interheaters are in the same furnace housing.

Effluent from the last reactor in the train is cooled to ambient temperature and led to a receiver in which the product mixture is separated into liquid and gas streams. Most of the separated gas stream, which is largely hydrogen, is compressed and recycled to the reactors, to provide the protective hydrogen partial pressure in the reaction environment. A net hydrogen product stream is withdrawn from the system by pressure control on the reaction system.

The receiver liquid, containing dissolved light hydrocarbons, is routed to a fractionator to produce a stabilized reformate suitable for blending into finished gasoline pools and generally free of hydrocarbons lighter than C_5.

Most variations in the flow-diagram arrangement of the several catalytic reforming processes in commercial use involve the concept of catalyst regeneration frequency. At relatively high pressures and high ratios of hydrogen recycle gas to naphtha feed, the catalyst is not rapidly fouled and deactivates rather slowly, so that continuous processing runs of a few months to more than a year are attainable. At the end of this uninterrupted processing, the reactor inlet-temperature requirements to maintain the desired reformate octane number may reach the limit of heater capabilities, or the catalyst selectivity may diminish to a point where it is economical to terminate the run and restore the activity and selectivity of the catalyst by *in situ* regeneration.

Regeneration is usually performed in place by burning the carbonaceous deposit accumulated on the catalyst with air diluted with combustion-product gases; the gas recycle compressor is used to circulate the mixture of air and combustion gas through the reactor system at a controlled burning temperature. This regeneration procedure, with proprietary additional steps, can restore the performance of the catalyst to what it was at the beginning of the preceding processing cycle. In an alternative procedure, which avoids the downtime involved in *in situ* regeneration, the spent catalyst is unloaded and the reactors are reloaded with fresh material. The spent catalyst is generally returned to the supplier for recovery of its precious-metal content.

At the other extreme is the concept of regeneration frequency in the process designed to operate at distinctly lower pressures and lower ratios of hydrogen recycle gas to naphtha. Higher reformate yields of a given octane number are obtainable; but, under these conditions, the catalyst fouls and deteriorates much more rapidly, so frequent catalyst regeneration is necessary. These plants are provided with an additional reactor, so manifolded to the other reactors and appropriately valved that any reactor can be taken off-line and regenerated, while the others continue to process naphtha feed. A reactor can be isolated from processing service for regeneration as often as once a day. Although higher reformate yields are obtainable with this "swing reactor" design, the units tend to be more expensive, because of the additional equipment required.

Between these extremes are other designs and operating techniques that can perform at any intermediate regeneration frequency; the choice of either extreme or an intermediate depends on the economics of the particular refining situation.

In the last 3 years, a new concept in catalytic reforming has been developed. It involves continuous catalyst regeneration and, by proper integration of the operation of the two sections, permits truly continuous reforming. These units are generally operated at lower pressures, to take advantage of the higher yields attained under such conditions, and usually have bimetallic catalysts. In this continuous separation, the UOP design involves a slow but continuous withdrawal of the catalyst from the lowest reactor in the stacked configuration; the catalyst is continuously regenerated in an external system and then returned to the top reactor. There are nine continuous platformers in operation and 40 in various stages of design and construction.

WORLDWIDE CATALYTIC REFORMING

The extent to which catalytic reforming is applied in the petroleum-refining industry is shown in Figure A-6 as daily capacity for the world and for each hemisphere for each year from 1960 to 1973. Capacities for the eastern hemisphere and the world exclude the U.S.S.R. and its bloc, for which data were not available. In those 14 years, the catalytic reforming capacity of the world increased by a factor of 2.5, having risen from 2.8 to 7 million barrels/day (445,000 to 1,110,000 m^3/day). Most of this increase has occurred in the eastern hemisphere, where the application of catalytic reforming has been catching up, so to speak, with that in the other hemisphere. Along with the general slowdown in overall refinery expansion since 1971, the reforming capacity (Figure A-6) has followed this trend in all sectors of the world.

Another way of regarding the application of catalytic reforming is by indicating its capacity as a percentage of crude oil processed in a given area (Figure A-7). About 50 million barrels (7,950,000 m^3) of crude oil were being processed per day early in 1973; therefore, the 7-million-barrel/day (1,110,000-m^3/day) worldwide catalytic reforming capacity

Appendix: Catalytic Reforming in Refining of Petroleum

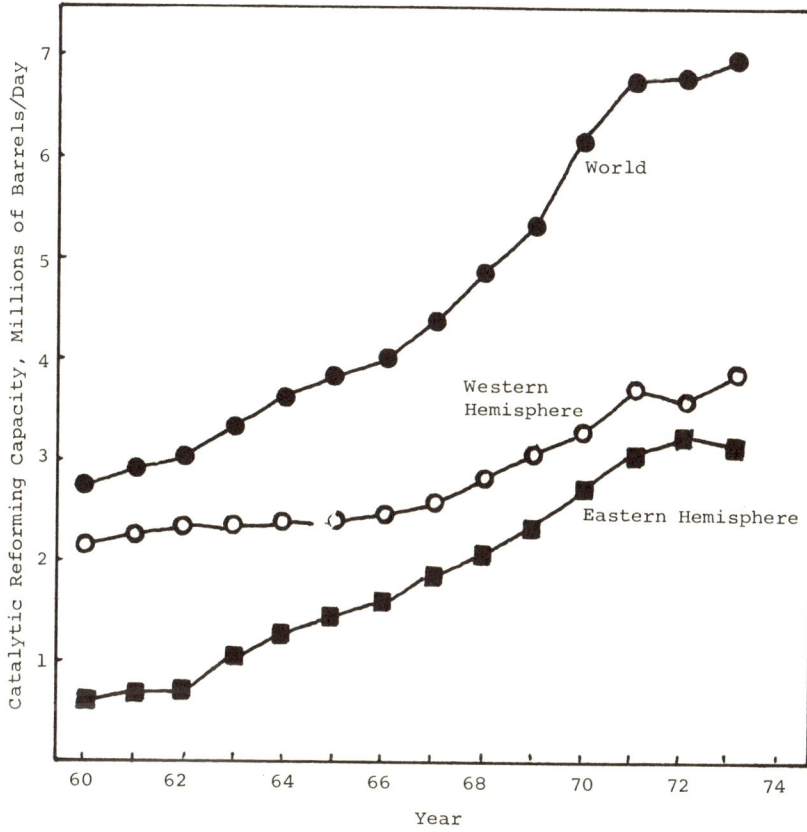

FIGURE A-6 Daily catalytic reforming capacity in petroleum refining. Data from Universal Oil Products Company.

represented about 14% of the crude oil. As shown in the figure, the percentage had not changed appreciably on a worldwide basis since 1960. The western hemisphere catalytically reforms a greater proportion (about 18% in recent years) of its processed crude oil than the eastern hemisphere (about 12%). As is also shown in Figure A-7, the proportion reformed in the western hemisphere had not changed notably since 1960, but there had been a sharp increase in the eastern hemisphere, from 8% to 13% (in 1970) and 12% (in 1971 and 1972).

The plot in Figure A-7 shows that the United States was reforming about 24% of its processed crude oil in 1973; this proportion represents very nearly the entire native C_7 400° F (200° C) content of the average crude oil refined in the United States. However, a small amount of

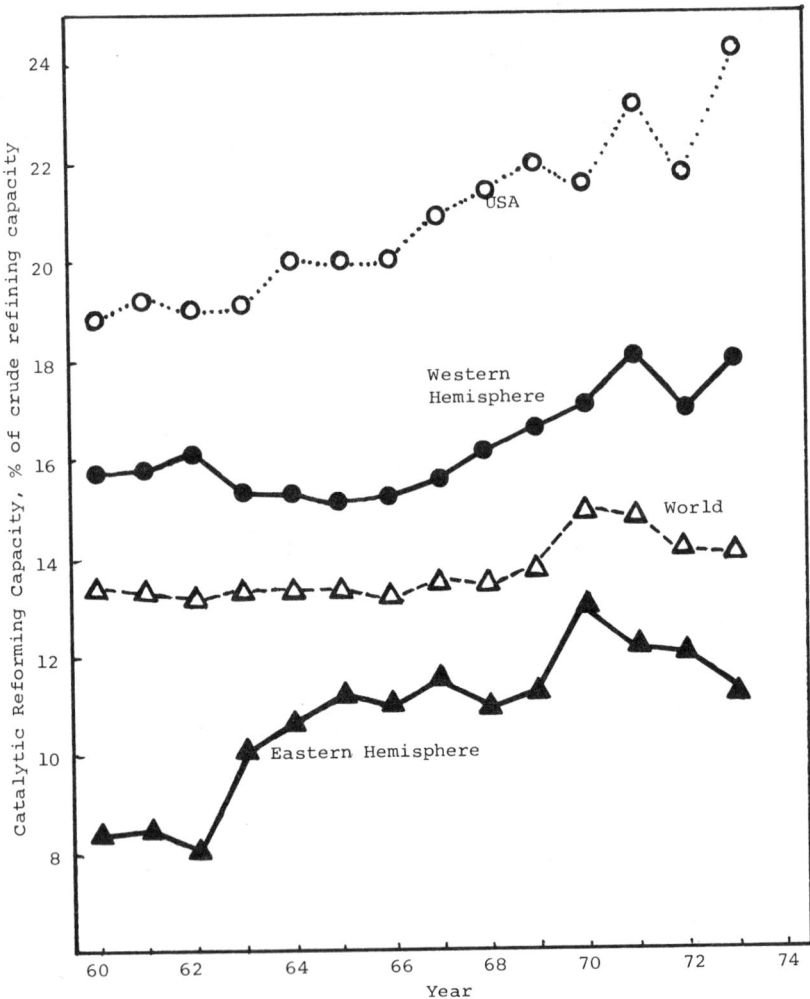

FIGURE A-7 Catalytic reforming capacity, as a percentage of crude-oil refining capacity. Data from Universal Oil Products Company.

U.S. reforming capacity is occupied with the processing of thermal and hydrocracked naphthas.

Although nearly all the available naphtha in western hemisphere crude oils is catalytically reformed, only about two-thirds of the estimated 19% of potentially reformable naphtha in the eastern hemisphere oils is being reformed. This suggests that there is a large potential for expanding catalytic reforming capacity in the eastern hemisphere.

Appendix: Catalytic Reforming in Refining of Petroleum

CONTRIBUTION OF CATALYTIC REFORMING TO THE U.S. GASOLINE POOL

The nation's gasoline pool is composed of five broad classes of gasolines, defined on the basis of their origin: straight-run, thermal, catalytically cracked, catalytically reformed, and alkylate and polymer. Although the national gasoline pool can be resolved into premium and regular grades as actually marketed (the proportions have changed with time), it is considered as a single commodity in this discussion. Even though there are several methods that could be used for characterizing the quality of gasoline, for simplicity only the research-method octane numbers of the pool and its components will be mentioned here.

The changing composition and octane numbers of the nation's gasoline pool are shown in Table A-2. In this table, the "unleaded" octane number of the pool represents the gasoline quality as produced by the refiner's processing units, and the "leaded" octane number represents what is dispensed at the filling station. Of interest is the dramatic increase of 25 in the octane number of the unleaded gasoline during the last three decades. This has been made possible by the increasing application of catalytic processing during that period. In 1940, catalytic processing was just beginning, and the nation's gasoline pool was essentially a blend of equal amounts of low-octane-number straight-run and thermally cracked gasoline having a moderate octane rating. By 1950, the pool gasoline had risen in unleaded octane number

TABLE A-2 Composition and Quality of U.S. Gasoline Pool[a]

	Year			
	1940	1950	1960	1972
Composition, vol%:				
Straight-run	50	40	19	12
Thermal	46	32	10	4
Catalytically cracked	2	20	31	38
Catalytically reformed	0	1	30	33
Alkylate and polymer	2	7	10	13
	100	100	100	100
Research octane number:				
Unleaded	64.3	75.3	86.2	89.3
Leaded	74.6	85.1	94.0	96.8
Lead content of pool, g/gal (U.S.)	1.5	2.2	2.0	2.3

[a] Data from Sterba.[428]

from 64 to 75, with the aid of catalytic cracking, alkylation, and polymerization.

As Table A-2 shows, the unleaded pool octane number rose by another 11 units in the next decade, from 1950 to 1960, largely through the extensive use of catalytic reforming. During that decade, the proportion of catalytic reformate in the nation's gasoline pool rose sharply from a nominal 1% to just over 30%, with much smaller increases in the percentages of alkylate and catalytically cracked gasolines. This represents an impressive contribution of catalytic reforming to improving the antiknock quality of the gasoline pool. Also to be noted in Table A-2 is the slight drop in the use of lead alkyls during this decade. This suggests that it was more economical to increase octane numbers by catalytic reforming than to use lead alkyls to achieve octane ratings required by the high-compression automobile engines.

During the 12-year period from 1960 to 1972, the octane ratings rose about 3 units, as shown in the table, with all catalytic processes contributing to the increase. As pointed out earlier, nearly all the available naphtha in the nation's processed crude oil is being catalytically reformed. However, catalytic reforming can continue to contribute to raising the antiknock rating of the nation's gasoline pool, because, by modifying the conditions of operation in the reforming reactors, the octane number of the reformate can be increased substantially. By contrast, the octane numbers of the gasoline products from other catalytic processes are relatively constant. Thus, the unleaded research octane number of catalytically cracked gasolines is in the range of 90–93, and that of alkylate and polymer gasoline ranges typically from 92 to 95. However, the catalytic reforming process is versatile in this respect. In the early 1950s, when the process was being commercialized, the national average unleaded octane number of reformates was just under 85; it is now in the vicinity of 95, and there are some sustained commercial operations producing reformates with octane numbers of over 100.

AROMATICS

In addition to its importance in helping to provide the world with high-octane-number motor fuels, the catalytic reforming process has been the primary instrument in the synthesis and supply of the basic aromatic building blocks—benzene, toluene, and the C_8 aromatics.

Of the total benzene consumed about 25 years ago, the portion made by the petroleum industry was around 5%; this figure has since risen to well over 90%. During this span of years, the production growth rate of

Appendix: Catalytic Reforming in Refining of Petroleum

each of these primary aromatics has been about 10%/year. These aromatics are produced in high purities by extraction of selected boiling-range catalytic reformates, largely with sulfolane and polyethylene glycol solvents.

USEFUL BYPRODUCTS

In the manufacture of high-quality motor fuel and aromatics, the simultaneous production of hydrogen should be considered a product, rather than a byproduct, of catalytic reforming, and it is generally profitable to maximize its yield. For the United States, this production of hydrogen from catalytic reforming amounts to an estimated 2.5×10^9 ft^3 (about 71×10^6 m^3) per day. This hydrogen is useful in the catalytic hydrotreatment of over 4 million barrels (about 635,000 m^3) of a wide variety of stocks per day. In addition, some of the hydrogen is used for the 750,000 barrels (120,000 m^3) required for hydrocracking per day.

Of the light hydrocarbons made in catalytic reforming, propane finds its way into liquid petroleum gas, isobutane goes to alkylation units, and n-butane is used to adjust the vapor pressure of finished gasoline. Only small amounts of methane and ethane are directed to refinery fuel.

References

1. Acres, G. J. K. Platinum catalysts for diesel engine exhaust purification. Platinum Met. Rev. 14:78–85, 1970.
2. Acres, G. J. K. Platinum catalysts for the control of air pollution: The elimination of organic fume by catalytic combustion. Platinum Met. Rev. 14:2–10, 1970.
3. Acres, G. J. K. The control of air pollution: Platinum catalyst systems for industrial odour control. Platinum Met. Rev. 15:9–12, 1971.
4. Acres, G. J. K., and B. J. Cooper. Automobile emission control systems: Platinum catalysts for exhaust purification. Platinum Met. Rev. 16:74–86, 1972.
5. Acres, G. J. K., B. J. Cooper, and G. L. Matlak. The production of automobile emission control catalysts: Global capability of the Johnson Matthey Group. Platinum Met. Rev. 17:82–87, 1973.
6. Acres, G. J. K., B. J. Cooper, E. Shutt, and B. W. Malerbi. Platinum catalysts for exhaust emission control: The mechanism of catalyst poisoning by lead and phosphorous compounds. Adv. Chem. Ser. 143:54–71, 1975.
7. Adriaenssens, E., and P. Knoop. A study of the optimal conditions for flameless atomic absorption spectrometry of iridium, platinum and rhodium. Anal. Chim. Acta 68:37–48, 1975.
8. Ahearn, A. J. Spark source mass spectrometric analysis of solids, pp. 347–384. In W. W. Meinke and B. F. Scribner, Eds. Trace Characterization. Chemical and Physical. National Bureau of Standards Monograph 100. Washington, D.C.: U.S. Government Printing Office, 1967.
9. Air quality criteria for carbon monoxide. Environment Reporter (Federal Laws Section 31):1951–1955, 1970.
10. Alexander, R. A., N. C. Baenziger, C. Carpenter, and J. R. Doyle. Metal-olefin compounds. I. The preparation and molecular structure of some metal-olefin compounds containing norbornadiene (bicyclo[2.2.1.]hepta-2,5-diene). J. Amer. Chem. Soc. 82:535–538, 1960.

References

11. Allen, A. D., F. Bottomley, R. O. Harris, V. P. Reinsalu, and C. V. Senoff. Ruthenium complexes containing molecular nitrogen. J. Amer. Chem. Soc. 89:5595–5599, 1967.
12. Allen, M. J. The use of platinum anodes in organic anodic processes. Platinum Met. Rev. 3:131–135, 1959.
13. Almar-Naess, A., and J. M. Drugli. Prevention of corrosion in paper making machines: Cathodic polarisation with platinum-titanium anodes. Platinum Met. Rev. 10:48–51, 1966.
14. American Bureau of Metal Statistics. [Tables on platinum metals], pp. 129–130. In Year Book of the American Bureau of Metal Statistics. Fifty-Third Annual Issue for the Year 1973. (Issued June 1974) New York: American Bureau of Metal Statistics, 1974.
15. American Conference of Governmental Industrial Hygienists. Threshold Limit Values for Chemical Substances and Physical Agents in Workroom Environment with Intended Changes for 1973. Cincinnati, Ohio: American Conference of Governmental Industrial Hygienists, 1973. 94 pp.
16. American Lava Corporation. (Technical Ceramic Products Division, 3M Company) ThermaComb® Brand Corrugated Ceramics. Bulletin No. 721. Chattanooga, Tenn.: American Lava Corporation, 1972. 6 pp.
17. American Society for Testing and Materials, Committee E-2 on Emission Spectroscopy. Methods for Emission Spectrochemical Analysis. (6th ed.) Philadelphia: American Society for Testing and Materials, 1971. 1,094 pp.
18. Amirnazmi, A., J. E. Benson, and M. Boudart. Oxygen inhibition in the decomposition of NO on metal oxides and platinum. J. Catalysis 30:55–65, 1973.
19. Analytical chemistry of the platinum-group metals. [Papers presented at] Symposium held at the National Institute for Metallurgy, Johannesburg, 2nd-4th February, 1972. J. S. Afr. Chem. Inst. 25:155–319, 1972.
20. Andersen, C. A., Ed. Microprobe Analysis. New York: John Wiley & Sons, 1973. 571 pp.
21. Anderson, J. N. Applied Dental Materials. (2nd ed.) Oxford: Blackwell Scientific Publications, 1961. 356 pp.
22. A new mining area for Rustenberg. Platinum Met. Rev. 18:64, 1974.
23. An improved titanium alloy for chemical plant: Palladium addition increases resistance to corrosion. Platinum Met. Rev. 3:88–89, 1959.
24. An organic process for the manufacture of hydrogen peroxide. Platinum Met. Rev. 3:54–55, 1959.
25. Armor, J. N., and H. Taube. Equilibria and rates in the formation of $[Ru(NH_3)_5N_2]^{2+}$ and $[(Ru(NH_3)_5)_2N_2]^{4+}$. J. Amer. Chem. Soc. 92:6170–6174, 1970.
26. Armor, J. N., and H. Taube. Reduction of nitrous oxide in the presence of pentaammineaquoruthenium(II). J. Amer. Chem. Soc. 93:6476–6480, 1971.
27. Augustine, R. L. Catalytic Hydrogenation. Techniques and Applications in Organic Synthesis. New York: Marcel Dekker, Inc., 1965. 188 pp.
28. Austin, T. C., R. B. Michael, and G. R. Service. Passenger Car Fuel Economy Trends Through 1976. Society of Automotive Engineers Technical Paper 750957 Presented at Automobile Engineering Meeting, Detroit, Michigan, October 13–17, 1975. 20 pp.
29. Bair, W. J., L. A. Temple, D. H. Willard, J. L. Terry, and A. Graybeal. Deposition and Retention of Ru^{106} Following Administration of $Ru^{106}O_2$ to Mice by Inhalation and Intratracheal Injection. Hanford Atomic Products Operation Report HW-52285. Richland, Wash.: General Electric Company, 1957. 24 pp.

References

30. Bair, W. J., D. H. Willard, and L. A. Temple. The behavior of inhaled $Ru^{106}O_2$ particles. Health Phys. 5:90–98, 1961.
31. Baird, M. C., C. J. Nyman, and G. Wilkinson. The decarbonylation of aldehydes by tris(triphenylphosphine)chlororhodium(1). J. Chem. Soc. A1968:348–351, 1968.
32. Baker, R. T. K., R. B. Thomas, and J. H. F. Notton. The behaviour of platinum catalysts for ammonia oxidation: Studies by controlled atmosphere microscopy. Platinum Met. Rev. 18:130–136, 1974.
33. Basolo, F., H. B. Gray, and R. G. Pearson. Mechanism of substitution reactions of complex ions. XVII. Rates of reaction of some platinum(II) and palladium(II) complexes with pyridine. J. Amer. Chem. Soc. 82:4200–4203, 1960.
34. Basolo, F., and R. G. Pearson. The *trans* effect of metal complexes. Prog. Inorg. Chem. 4:381–453, 1962.
35. Beamish, F. E. The Analytical Chemistry of the Noble Metals. New York: Pergamon Press, 1966. 608 pp.
36. Beamish, F. E., W. A. E. McBryde, and R. R. Barefoot. The platinum metals, pp. 304–335. In C. A. Hampel, Ed. Rare Metals Handbook. (2nd ed.) New York: Reinhold Publishing Corporation, 1961.
37. Beamish, F. E., and J. C. Van Loon. Recent Advances in the Analytical Chemistry of the Noble Metals. New York: Pergamon Press, 1972. 511 pp.
38. Beath, C. B., R. J. Westwood, and C. A. Cousins. Platinum mining at Rustenburg. The development of operating methods. Platinum Met. Rev. 5:102–108, 1961.
39. Beattie, R. W. Palladium plating on telephone plugs and sockets. Platinum Met. Rev. 6:52–56, 1962.
40. Beck, W., E. Schuierer, and K. Feldl. New metal-azide complexes. Angew. Chem. Int. Ed. 5:247, 1966.
41. Bell, B. H. J. Platinum catalysts in ammonia oxidation: Operating conditions in Fisons new nitric acid plant. Platinum Met. Rev. 4:122–126, 1960.
42. Belluco, U., R. Ettorre, F. Basolo, R. G. Pearson, and A. Turco. Activation parameters for some substitution reactions of acidodiethylenetriamineplatinum(II) complexes. Inorg. Chem. 5:591–593, 1966.
43. Bennett, H. E. The Pallador thermocouple. Platinum Met. Rev. 4:66–67, 1960.
44. Bentham, J. E., S. Cradock, and E. A. V. Ebsworth. Silyl and germyl compounds of platinum and palladium. Part I. Platinum derivatives of monosilane and monogermane. J. Chem. Soc. A1971:587–593.
45. Bentley, D. R., and D. J. Schweibold. Questor reverter emission control system total vehicle concept. SAE (Soc. Automot. Eng.) Trans. 82:852–874, 1973.
46. Berta, D. A., W. A. Spofford, P. Boldrini, and E. L. Amma. The crystal and molecular structure of tetrakis((thiourea)palladium(II) chloride. Inorg. Chem. 9:136–142, 1970.
47. Bertodo, R. High temperature strain gauges for turbo-jet components: Advantages of platinum alloy resistance wires. Platinum Met. Rev. 8:128–130, 1964.
48. Betteridge, W., and J. Hope. The separation of hydrogen from gas mixtures: A process of absorption and desorption by palladium. Platinum Met. Rev. 19:50–59, 1975.
49. Blair, J., and J. G. Gibb. Efficiency of platinum gauzes in the manufacture of nitric acid: A method for determining the frequency of pickling. Platinum Met. Rev. 11:100–103, 1969.
50. Blum, J., J. Y. Becker, H. Rosenman, and E. D. Bergmann. Homogeneous benzylic oxidation catalyzed by some complexes of the platinum group. J. Chem. Soc. B1969:1000–1004.

References

51. Bond, G. C. Catalysis by Metals. New York: Academic Press, 1962. 519 pp.
52. Bond, G. C. Platinum metal salts and complexes as homogeneous catalysts: Scope for novel chemical processes. Platinum Met. Rev. 8:92–97, 1964.
53. Bond, G. C. Platinum metals as hydrogenation catalysts. Platinum Met. Rev. 1:87–93, 1957.
54. Bond, G. C., and G. Webb. Ruthenium and osmium as hydrogenation catalysts. Platinum Met. Rev. 6:12–19, 1962.
55. Bond, G. C., and D. E. Webster. Ruthenium-platinum oxide catalysts for hydrogenation reactions: A critical comparison of published results. Platinum Met. Rev. 13:57–60, 1969.
56. Bond, G. C., and D. E. Webster. Ruthenium-platinum oxide catalysts: High activity in hydrogenation reactions. Platinum Met. Rev. 9:12–13, 1965.
57. Booth, G. Complexes of the transition metals with phosphines, arsines, and stibines. Adv. Inorg. Chem. Radiochem. 6:1–69, 1964.
58. Boumans, P. W. J., and F. J. de Boer. An assessment of the inductively coupled high-frequency plasma for simultaneous multi-element analysis. Proc. Anal. Div. Chem. Soc. (Lond.) 12:140–152, 1975.
59. Bowen, H. J. M. Trace Elements in Biochemistry. New York: Academic Press, 1966. 241 pp.
60. Bromfield, R. J., R. H. Dainty, R. D. Gillard, and B. T. Heaton. Growth of microorganisms in the presence of transition metal complexes: The antibacterial activity of $trans$-dihalogenotetrapyridinerhodium(III) salts. Nature 223:735–736, 1969.
61. Brooks, E. H., and R. J. Cross. Group IVB metal derivatives of the transition elements. Organomet. Chem. Rev. A6:227–282, 1970.
62. Brown, C. K., and G. Wilkinson. Homogenous hydroformylation of alkenes with hydrido carbonyltris-(triphenylphosphine)rhodium(I) as catalyst. J. Chem. Soc. A1970:2753–2764.
63. Brown, K. W. Ruthenium: Its Behavior in Plant and Soil Systems. EPA-600/3-76-019. Las Vegas, Nev.: U.S. Environmental Protection Agency, Environmental Monitoring & Support Laboratory, 1976. 28 pp.
64. Browning, E. Toxicity of Industrial Metals. London: Butterworths, 1961. 339 pp.
65. Browning, E. Toxicity of Industrial Metals. (2nd ed.) London: Butterworths, 1969, p. 262.
66. Brubaker, P. E., J. P. Moran, K. Bridbord, and F. G. Heuter. Noble metals: A toxicological appraisal of potential new environmental contaminants. Environ. Health Perspect. 10:39–56, 1975.
67. Bruckner, H. W., C. C. Cohen, G. Deppe, B. Kabakow, R. C. Wallach, E. M. Greenspan, S. B. Gusberg, and J. F. Holland. Chemotherapy of gynecological tumors with platinum II. J. Clin. Hematol. Oncol. (Wadley Institutes of Molecular Medicine) 7:619–632, 1977.
68. Bruckner, H. W., C. J. Cohen, S. B. Gusberg, R. C. Wallach, B. Kabakow, E. M. Greenspan, and J. F. Holland. Chemotherapy of ovarian cancer with adriamycin (ADM) and cis-platinum (DDP). Abstract C-204. In Proceedings of the AACR/ASCO (American Association for Cancer Research/American Society of Clinical Oncology) Meeting, Montreal, Canada, May 1976.
69. Bruner, H. D. Distribution of Ru^{106} in rats after intravenous injection of $Ru^{106}O_2 \cdot 3H_2O$. Radiat. Res. 5:471–472, 1956. (abstract)
70. Burbage, J. J., and W. C. Fernelius. Reduction of potassium cyanopalladate(II) by potassium in liquid ammonia; a zerovalent compound of palladium. J. Amer. Chem. Soc. 65:1484–1486, 1943.

References

71. Burke, D. P. Catalysts. Part 1: Petroleum catalysts. A comprehensive look at a $168-million/year business headed for spectacular growth. Chem. Week 111(18):23–33, 1972.
72. Butterman, W. C. Platinum-group metals, pp. 1037–1049. In U.S. Department of the Interior, Bureau of Mines. Minerals Yearbook 1973. Vol. I. Metals, Minerals, and Fuels. Washington, D.C.: U.S. Government Printing Office. 1975.
73. Butterman, W. C. Platinum-group metals. Preprint, 12 pp. In Bureau of Mines Minerals Yearbook 1975. Vol. 1. Washington, D.C.: U.S. Government Printing Office. 1977.
74. Cadle, S. H., D. P. Chock, J. M. Heuss, and P. R. Monson. Results of the General Motors Sulfate Dispersion Experiment. Environmental Science Department Research Publication GMR-2107. Warren, Mich.: General Motors Corporation Research Laboratories, 1976. [249 pp.]
75. Cajka, C. J. Platinum metals, pp. 310–315. In Canadian Minerals Yearbook 1971. Mineral Report 21. Ottawa: Department of Energy, Mines and Resources, Mineral Resources Branch, 1973.
76. Calvin, G., and G. E. Coates. Organopalladium compounds. J. Chem. Soc. 1960:2008–2016.
77. Campbell, K. I., E. L. George, L. L. Hall, and J. F. Stara. Dermal irritancy of metal compounds. Arch. Environ. Health 30:168–170, 1975.
78. Canterford, J. H., and R. Colton. [The platinum group metals], pp. 322–389. In Halides of the Transition Elements. Vol. 3. Halides of the Second and Third Row Transition Metals. New York: John Wiley & Sons, Ltd., 1968.
79. Casarett, L. J., S. Bless, R. Katz, and J. K. Scott. Retention and fate of iridium-192 in rats following inhalation. Amer. Ind. Hyg. Assoc. J. 21:414–418, 1960.
80. Catalytic heaters called potential threat to life unless vented. Environ. Health Lett. 14(2):5, 1975.
81. Cattalini, L., and M. Martelli. Relazione tra reattivita di complessi planari del platino e natura del gruppo uscente, in reazioni di sostituzione nucleofila. Gaz. Chim. Ital. 97:498–508, 1967.
82. C[haston], J. C. High temperature resistance furnaces: 40 per cent rhodium-platinum versus rhodium. Platinum Met. Rev. 8:66, 1964.
83. C[haston], J. C. Organic deposits on noble metal contacts: An investigation into contact contamination in telephone relays. Platinum Met. Rev. 3:19–21, 1959.
84. C[haston], J. C. Palladium telephone contacts: A new series of miniature wire spring relays. Platinum Met. Rev. 12:14–15, 1968.
85. Chaston, J. C. Reactions of oxygen with the platinum metals: I–The oxidation of platinum. Platinum Met. Rev. 8:50–54, 1964.
86. Chaston, J. C. Reactions of oxygen with the platinum metals: II–Oxidation of ruthenium, rhodium, iridium, and osmium. Platinum Met. Rev. 9:51–56, 1965.
87. Chaston, J. C. Reactions of oxygen with the platinum metals: IV–The oxidation of palladium. Platinum Met. Rev. 9:126–219, 1965.
88. Chatt, J. Hydride complexes. Science 160:723–729, 1968.
89. Chatt, J. The nature of the co-ordinate link. Part I. The non-ionic complex compounds of tri-n-propylphosphine with platinic and platinous chlorides. J. Chem. Soc. 1950:2301–2310.
90. Chatt, J., R. S. Coffey, A. Gough, and D. T. Thompson. The reversible reaction between olefins and platinum hydrides. J. Chem. Soc. A1968:190–194.
91. Chatt, J., L. A. Duncanson, and B. L. Shaw. A volatile chlorohydride of platinum. Proc. Chem. Soc. (Lond.) 1957:343.

92. Chatt, J., L. A. Duncanson, and B. L. Shaw. The influence of ligands on the Pt-H stretching frequency in a series of complex hydrides of platinum(II); a complex hydride of platinum. Chem. Ind. (Lond.) 1958:859–860.
93. Chatt, J., C. Eaborn, S. D. Ibekwe, and P. N. Kapoor. Preparation and properties of compounds containing platinum-silicon bonds. J. Chem. Soc. A1970:1343–1351.
94. Chatt, J., G. J. Leigh, and A. P. Storace. Complexes of ruthenium halides with organic sulphides (thioethers). J. Chem. Soc. A1971:1380–1389.
95. Chatt, J., and B. L. Shaw. Hydrido-complexes of platinum(II). J. Chem. Soc. 1962:5075–5084.
96. Chernyaev, I. I. The mononitrites of bivalent platinum. Ann. Inst. Platine USSR 4:243–275, 1926. (in Russian)
97. Cleare, M. J. Chemistry of co-ordination complexes. Recent Results Cancer Res. 48:12–37, 1974.
98. Cleare, M. J. Coordination compounds of the platinum group metals: A review of their preparative methods and applications. Platinum Met. Rev. 18:122–129, 1974.
99. Cleare, M. J., E. G. Hughes, B. Jacoby, and J. Pepys. Immediate (type 1) allergic responses to platinum compounds. Clin. Allergy 6:183–195, 1976.
100. Clements, F. S. Twenty-five years' progress in platinum metals refining. Ind. Chemist 38:345–354, 1962.
101. Cockayne, B. Czochralski growth of oxide single crystals: Iridium crucibles and their use. Platinum Met. Rev. 18:86–91, 1974.
102. Connor, H. A laboratory scale hydrogen purification unit. Platinum Met. Rev. 9:7–8, 1965.
103. C[onnor], H. Palladium chloride catalyst in olefin oxidations: New production processes for acetone and methyl ethyl ketone. Platinum Met. Rev. 7:132–133, 1963.
104. C[onnor], H. Production of nitric acid: Catalyst costs in modern processes. Platinum Met. Rev. 14:61, 1970.
105. Connor, H. The manufacture of nitric acid: The role of platinum alloy gauzes in the ammonia oxidation process. Platinum Met. Rev. 11:2–9, 1967.
106. C[onnor], H. The synthesis of high molecular weight polymethylenes: New ruthenium catalysts of exceptional activity. Platinum Met. Rev. 7:105, 1963.
107. Connors, T. A., and J. J. Roberts, Eds. Platinum coordination complexes in cancer chemotherapy. Recent Results Cancer Res. 48:1–195, 1974.
108. Coombs, R. R. A., and P. G. H. Gell. Classification of allergic reactions responsible for clinical hypersensitivity and disease, pp. 761–781. In P. G. H. Gell, R. R. A. Coombs and P. J. Lachmann, Eds. Clinical Aspects of Immunology. (3rd ed.) Oxford: Blackwell Publications, 1975.
109. Cotton, F. A., and G. Wilkinson. [The platinum-group metals], pp. 990–1044. In Advanced Inorganic Chemistry. A Comprehensive Text. (3rd ed.) New York: Interscience Publishers, 1972.
110. Cotton, J. B. Platinum-faced titanium for electrochemical anodes: A new electrode material for impressed current cathodic protection. Platinum Met. Rev. 2:45–57, 1958.
111. Cotton, J. B. The role of palladium in enhancing corrosion resistance of titanium. Platinum Met. Rev. 11:50–52, 1967.
112. Cousins, C. A. The Bushveld igneous complex: The geology of South Africa's platinum resources. Platinum Met. Rev. 3:94–99, 1959.
113. Crockett, J. H. Platinum, Sections B-G, K, M, O. In H. H. Wedepohl, Ed. Handbook of Geochemistry. Vol. II. New York: Springer-Verlag, 1969.

References

114. Cross, R. J. σ-Complexes of platinum(II) with hydrogen, carbon and other elements of group IV. Chem. Rev. 2:97–140, 1967.
115. Curry, S. W. Platinum catalysts in petroleum refining. Platinum Met. Rev. 1:38–43, 1957.
116. Cvitkovic, E., D. Hayes, and R. Golbey. Primary combination chemotherapy (VAB III) for metastatic or unresectable germ cell tumors. Proc. Amer. Assoc. Cancer Res. Amer. Soc. Clin. Oncol. 17:296, 1976. (abstract)
117. Darling, A. S. Thermal and electric palladium alloy diffusion cells: Complementary methods of obtaining ultra-pure hydrogen. Platinum Met. Rev. 7:126–129, 1963.
118. Dawson, G. W. Chemical Toxicity of Elements. BNWL-1815. Richland, Wash.: Battelle Pacific Northwest Laboratories, 1974. 25 pp.
119. Dean, L. E., H. R. Harris, D. H. Belden, and V. Haensel. The Penex process for pentane isomerisation. Platinum Met. Rev. 3:9–11, 1959.
120. Deluca, J. P., L. L. Murrell, R. P. Rhodes, and S. J. Tauster. The stabilization of ruthenium on MgO, Abstract COLL 38. In Abstracts of Papers. 170th National Meeting. American Chemical Society, Chicago, Illinois, August 24–29, 1975.
121. Dennis, W. H. Metallurgy of the Non-Ferrous Metals. London: Sir Isaac Pitman & Sons, 1954. 647 pp.
122. Dickens, P. G., R. Heckingbottom, and J. W. Linnett. Oxidation of metals and alloys. Part 2. Oxidation of metals by atomic and molecular oxygen. Trans. Faraday Soc. 65:2235–2247, 1969.
123. Dickerson, R. E., D. Eisenberg, J. Varnum, and M. L. Kopka. $PtCl_4^{2-}$: A methionine-specific label for protein crystallography. J. Mol. Biol. 45:77–84, 1969.
124. Doelp, L. C., R. W. Johnston, S. Gussow, and J. H. Olson. A Model for Roadside Dispersion of H_2SO_4. Marcus Hook, Penn.: Air Products and Chemicals, Inc., Houdry Division, 1976.
125. Duffield, F. V. P., A. Yoakum, J. Bumgarner, and J. Moran. Determination of human body burden baseline data of platinum through autopsy tissue analysis. Environ. Health Perspect. 15:131–134, 1976.
126. Durbin, P. W., K. G. Scott, and J. G. Hamilton. The distribution of radioisotopes of some heavy metals in the rat. Univ. Calif. Publ. Pharmacol. 3:1–34, 1957.
127. Early, J. E., and T. Fealey. Hydroxide ion as a reducing agent for mixed-valence ruthenium trimers. J. Chem. Soc. (Lond.) D1971:331.
128. Earwicker, G. A. The sulphito-compounds of palladium(II). J. Chem. Soc. 1960:2620–2626.
129. Edwards, J. H. A quantitative study on the activation of the alternative pathway of complement by mouldy hay dust and thermophilic actinomycetes. Clin. Allergy 6:19–25, 1976.
130. Edwards, R. I. The Refining of the Platinum-Group Metals. TMS Paper Selection No. A75–59. New York: The Metallurgical Society of AIME, 1975. 21 pp.
131. Einhorn, L. H., and B. Furnas. Improved chemotherapy in disseminated testicular cancer. J. Clin. Hematol. Oncol. (Wadley Institutes of Molecular Medicine) 7:662–671, 1977.
132. Einhorn, L. H., B. E. Furnas, and N. Powell. Combination chemotherapy of disseminated testicular carcinoma with cis-platinumdiammìnedichloride (CPDD), vinblastine (VLB), and bleomycin (BLEO). Abstract C-13. In Proceedings of the AACR/ASCO (American Association for Cancer Research/American Society of Clinical Oncology) Meeting, Montreal, Canada, May, 1976.
133. Endter, F. A high-temperature reactor for the synthesis of hydrogen cyanide. Platinum Met. Rev. 6:9–10, 1962.
134. Energy and Environmental Analysis, Inc. An Analysis of the Automotive Sulfate Question. Summary of Findings. (Prepared for the Manufacturers of Emission Con-

trols Association) Arlington, Va.: Energy and Environmental Analysis, Inc., [1975]. 29 pp.
135. Enomoto, Y., K. Watari, and R. Ichikawa. Studies on the metabolism of some chemical species of radio-ruthenium in the rat. 1. Early fate of ingested ruthenium. J. Radiat. Res. (Tokyo) 13:193-198, 1972.
136. Enoto, H., and H. Hayashi. A Study of the Exhaust Gas Cleaning Catalyst Converter Performance Test Method (No. 7)—Evaluation of the Hydrocarbon Oxidizing Power of a Monolithic Catalyst by Test Piece. Paper 8, Presented at 27th Study Meeting of the Society of Automotive Engineers of Japan, Tokyo, May 23-24, 1973. (in Japanese)
137. Erck, A., L. Rainen, J. Whileyman, I-M. Chang, A. P. Kimball, and J. Bear. Studies of rhodium(II) carboxylates as potential antitumor agents. Proc. Soc. Exp. Biol. Med. 145:1278-1283, 1974.
138. Everett, G. L. The determination of precious metals by flameless atomic-absorption spectrophotometry. Analyst 101:348-355, 1976.
139. Fassel, V. A., and R. N. Kniseley. Inductively coupled plasma—Optical emission spectroscopy. Anal. Chem. 46:1110A, 1111A, 1116A-1118A, 1120A, 1974.
140. Fassel, V. A., and R. N. Kniseley. Inductively coupled plasmas. Anal. Chem. 46:1155A, 1158A, 1162A, 1164A, 1974.
141. Feber, R. C. The removal of radioactive zirconium and ruthenium from process solutions: Permanganate pre-treatment. Prog. Nucl. Energy Ser. 3 2:247-256, 1958.
142. Fedor, J. R., C. H. Lee, and M. P. Makowski. Metallic Catalysts—An Approach to Achieving 1976 Emission Standards. Paper 37B Presented at the 74th National Meeting of the American Institute of Chemical Engineers, New Orleans, March 11-15, 1973. 27 pp.
143. Finklea, J. F., W. C. Nelson, J. B. Moran, G. G. Akland, R. I. Larsen, D. I. Hammer, and J. H. Knelson. Estimates of the Public Health Benefits and Risks Attributable to Equipping Light Duty Motor Vehicles with Oxidation Catalysts. Research Triangle Park, N.C.: National Environmental Research Center, U.S. Environmental Protection Agency, 1975. 71 pp.
144. Fisher, R. F., D. J. Holbrook, Jr., H. B. Leake, and P. E. Brubaker. Effect of platinum and palladium salts on thymidine incorporation into DNA of rat tissues. Environ. Health Perspect. 12:57-62, 1975.
145. Flindt, M. L. H. Pulmonary disease due to inhalation of derivatives of *Bacillus subtilis* containing proteolytic enzyme. Lancet 1:1177-1181, 1969.
146. Ford, L. A. Platinum alloy permanent magnets: The design of magnetic circuits for Platinax II. Platinum Met. Rev. 8:82-90, 1964.
147. Fowle, M. J. Platinum and the petroleum industry: A forecast of probable requirements. Platinum Met. Rev. 1:129-131, 1957.
148. Freedman, S. O., and J. Krupey. Respiratory allergy caused by platinum salts. J. Allergy 42:233-237, 1968.
149. Friedman, M. E., B. Musgrove, K. Lee, and J. E. Teggins. Inhibition of malate dehydrogenase by platinum(II) complexes. Biochim. Biophys. Acta 250:286-296, 1971.
150. Friedman, M. E., and J. E. Teggins. The reactivities of isomers of dichlorodiammine-platinum(II) with dehydrogenase enzymes: Evidence for inhibition via cross-linkage. Biochim. Biophys. Acta 350:263-272, 1974.
151. Fuel cells, pp. 614-621. In C. A. Hampel, Ed. The Encyclopedia of Electrochemistry. New York: Reinhold Publishing Corporation, 1964.
152. Fujita, S. Experimental studies on carcinogenicity of physical stimuli. Report 2. Carcinogenicity of Ag-Pd-Au alloy and acid mucopoly-saccharides in tumor in-

duced by this agent. Shika Igaku (J. Osaka Odontolog. Soc.) 34:918–932, 1971. (in Japanese, summary in English)
153. Fume elimination in enamelling ovens. Platinum Met. Rev. 1:50, 1957.
154. Furchner, J. E., C. R. Richmond, and G. A. Drake. Comparative metabolism of radionuclides in mammals—VII. Retention of ^{106}Ru in the mouse, rat, monkey and dog. Health Phys. 21:355–365, 1971.
155. Furlong, L. E., E. L. Holt, and L. S. Bernstein. Emission control and fuel economy, Abstract FUEL 4. In Abstracts of Papers. 167th National Meeting. American Chemical Society, Los Angeles, California, March 31–April 5, 1974.
156. Furukawa, G. T., M. L. Reilly, and J. S. Gallagher. Critical analysis of heat-capacity data and evaluation of thermodynamic properties of ruthenium, rhodium, palladium, iridium, and platinum from 0 to 300 K. A survey of the literature data on osmium. J. Phys. Chem. Ref. Data 3:163–209, 1974.
157. Gel'man, A. D., E. F. Karandashova, and L. N. Essen. Synthesis of the three stereoisomers of [Pt(C_5H_5N) –(NH_3)ClBr]. Dokl. Akad. Nauk. S.S.S.R. 63:37–40, 1948. (in Russian)
158. General Motors Corporation. Standard bench test evaluation, base metal vs noble metal, Fig. 3. In General Motors Request for Suspension of 1975 Federal Emissions Standards. Vol. 1. Section 4. System Performance. Warren, Mich.: General Motors Corporation, April 3, 1972.
159. General Motors Corporation. Statement of General Motors Corporation Before the Environmental Protection Agency at its Hearings on California's Request for Waiver of Federal Preemption with Respect to 1977 Model Year Light Duty Vehicles and Light Duty Trucks (Presented by E. S. Starkman, Vice President, Environmental Activities Staff), Los Angeles, California, April 29, 1975. Warren, Mich.: General Motors Corporation, 1976. 11 pp.
160. Glover, B. M. Growth of platinum reforming in Western Europe: Meeting the demand for high-octane fuels and aromatic hydrocarbons. Platinum Met. Rev. 6:86–91, 1962.
161. Gofman, J. W., O. F. deLalla, E. L. Kovich, O. Lowe, W. Martin, D. L. Piluso, R. K. Tandy, and F. Upham. Chemical elements of the blood of man. Arch. Environ. Health 8:105–109, 1964.
162. Goldberg, R. N., and L. G. Hepler. Thermochemistry and oxidation potentials of the platinum group metals and their compounds. Chem. Rev. 68: 229–252, 1968.
163. Goldschmidt, V. M. (Alex Muir, Ed.) Geochemistry. London: Oxford University Press, 1954. 730 pp.
164. Golovnya, V. A., and C. C. Ni. Oxidation of dinitrile platinum complexes. Zh. Neorg. Khim. 3:1954–1958, 1959. (in Russian)
165. Gottlieb, J. A., and B. Drewinko. Review of the current clinical status of platinum coordination complexes in cancer chemotherapy. Cancer Chemother. Rep. (Part 1) 59:621–628, 1975.
166. Gouldsmith, A. F. S., and B. Wilson. Extraction and refining of the platinum metals: A complex cycle of smelting, electronic and chemical operations. Platinum Met. Rev. 7:136–143, 1963.
167. Graves, C. Prevention of contact contamination in sealed relays. Platinum Met. Rev. 3:22–23, 1959.
168. Greenberg, M., J. F. Milne, and A. Watt. Survey of workers exposed to dusts containing derivatives of *Bacillus subtilis*. Brit. Med. J. 2:629–633, 1970.
169. Griffith, W. P. Osmium tetroxide and its applications. Platinum Met. Rev. 18:94–96, 1974.
170. Griffith, W. P. The Chemistry of the Rarer Platinum Metals. (Os, Ru, Ir and Rh) New York: Interscience Publishers, 1967. 491 pp.

171. Griffith, W. P., J. Lewis, and G. Wilkinson. Studies on transition metal nitric oxide complexes. Part V. Nitric oxide complexes of tetrahedral bivalent nickel and some other metals. J. Chem. Soc. 1959: 1775–1779.
172. Gross, G. P. The Effect of Fuel and Vehicle Variables on Polynuclear Aromatic Hydrocarbon and Phenol Emissions. SAE (Soc. Automot. Eng.) Trans. 81:830–855, 1972.
173. Guthrie, R. W., P. Melius, and J. E. Teggins. Inhibition of leucine aminopeptidase by halide complexes of platinum. J. Med. Chem. 14:75–76, 1971.
174. Gutt, W., and B. Hinkins. Design of rhodium platinum furnace elements: Long life at high temperature. Platinum Met. Rev. 12:86–88, 1968.
175. Habu, T. Histopathological effects of silver-palladium-gold alloy implantation on the oral submucous membranes and other organs. Shika Igaku (J. Osaka Odontolog. Soc.) 31(1):17–48, 1968. (in Japanese, summary in English)
176. Haensel, V., and H. S. Bloch. Duofunctional platinum catalysts in the petroleum industry. Platinum Met. Rev. 8:2–8, 1964.
177. Haensel, V., and M. J. Sterba. Comparison of platforming and thermal reforming. Adv. Chem. Ser. 5:60–75, 1951.
178. Hamilton, J. G. The metabolism of the fission products and of the heaviest elements. Radiology 49:325–343, 1947.
179. Hamilton, J. G. The metabolism of the radioactive elements created by nuclear fission. New Engl. J. Med. 240:863–870, 1949.
180. Hanson, W. C., and R. L. Browning. Absorption and distribution of ruthenium in fowl, pp. 95–98. In Biology Research Annual Report 1953 by the Staff of the Biology Section, January 4, 1954. HW-30437. Richland, Wash.: Hanford Atomic Products Operation, 1954.
181. Hara, M., K. Ohno, and J. Tsuji. Palladium-catalysed hydrosilation of olefins and polyenes. J. Chem. Soc. D 1971:247.
182. Harbord, N. N. Ammonia oxidation catalysts: Deposits on some rhodium-platinum gauzes. Platinum Met. Rev. 18:97–102, 1974.
183. Harder, H. C. Effects of platinum compounds on bacteria, viruses and cells in culture. Recent Results Cancer Res. 48:98–111, 1974.
184. Harder, H. C., and B. Rosenberg. Inhibitory effects of anti-tumor platinum compounds on DNA, RNA and protein synthesis in mammalian cells *in vitro*. Int. J. Cancer 6:207–216, 1970.
185. Hardison, L. C. A summary of the use of catalysts for stationary emission source control, pp. 271–296. In B. R. Banerjee, Ed. Proceedings of the First National Symposium on Heterogeneous Catalysis for Control of Air Pollution, Philadelphia, Nov. 21–22, 1968. Durham, N.C.: U.S. Public Health Service, National Air Pollution Control Administration, [1968.]
186. Hartley, F. R. The Chemistry of Platinum and Palladium. New York: John Wiley & Sons, 1973. 544 pp.
187. Hatanaka, M., R. Mahakawa, and H. Maruyama. Flame Retardant Silicone Rubber Compositions. U.S. Patent 3,862,082, Jan. 21, 1975. 7 pp.
188. Hawes, M. G. Preparation of heavy water by catalytic exchange. Platinum Met. Rev. 3:118–124, 1959.
189. Hawkins, D. T., and R. Hultgren. Constitution of binary alloys, pp. 251–367. In T. Lyman, Ed. Metals Handbook. (8th ed.) Vol. 8. Metallography, Structures and Phase Diagrams. Metals Park, Ohio: American Society for Metals, 1973.
190. Hawley, J. E. The Sudbury ores: Their mineralogy and origin. Can. Mineralogist 7:1–207, 1962.

References

191. Hearle, J. W. S., and A. Johnson. Platinum alloys in the production of viscose rayon: The selection of materials for spinning jets. Platinum Met. Rev. 5:2-8, 1961.
192. Hébert, R. Affections porvoquées par les composés du platine. Arch. Mal. Prof. 27:877-886, 1966.
193. Hegedus, L. L. Effects of channel geometry on the performance of catalytic monoliths, Abstract PETR 008. In Abstracts of Papers. 166th National Meeting, American Chemical Society, Chicago, Illinois, August 26-31, 1973.
194. Heywood, A. E. Platinum recovery in ammonia oxidation plants: Experience of the gold-palladium catchment gauze system. Platinum Met. Rev. 17:118-129, 1973.
195. High temperature strain gauges. Platinum Met. Rev. 7:53, 1963.
196. Hightower, J. W. Catalysts for automobile emission control, pp. 615-636. In D. Delmon, P. A. Jacobs, and G. Poncelet, Eds. Preparation of Catalysts. Scientific Bases for the Preparation of Heterogeneous Catalysts. Proceedings of the International Symposium, Brussels, 1975. Amsterdam: Elsevier Scientific Publishing Company, 1976.
197. Hightower, J. W. Statement of Joe W. Hightower, Department of Chemical Engineering, Rice University, pp. 52-65. In Research and Development Related to Sulfates in the Atmosphere. Hearings Before the Subcommittee on the Environment and the Atmosphere of the Committee on Science and Technology. U.S. House of Representatives. Ninety-fourth Congress First Session, July 8, 9, 11, 14, 1975. [No. 39] Washington, D.C.: U.S. Government Printing Office, 1976.
198. Hill, J. The anodic protection of steel: A review of recent progress. Platinum Met. Rev. 7:94-95, 1963.
199. Hill, J. M., E. Loeb, A. MacLellan, N. O. Hill, A. Khan, and J. J. King. Clinical studies of platinum coordination compounds in the treatment of various malignant diseases. Cancer Chemother. Rep. I 59:647-659, 1975.
200. Hirao, O. Automobiles and air pollution. Technocrat 8(8):15-20, 1975.
201. Hoar, T. P. Corrosion resistance of chromium: Effects of additions of platinum metals. Platinum Met. Rev. 5:141-143, 1961.
202. H[oar], T. P. Increasing the acid resistance of stainless steels: Influence of additions of platinum metals. Platinum Met. Rev. 2:117-119, 1958.
203. Hoar, T. P. Increasing the resistance of titanium to non-oxidising acids. Platinum Met. Rev. 4:59-64, 1960.
204. Hofmeister, F. Ueber die physiologische Wirking der Platinbasen. Naunyn-Schmiedeberg's Arch. Exp. Path. Pharmakol. 16:393-439, 1883.
205. Holbrook, D. J., Jr., M. E. Washington, H. B. Leake, and P. E. Brubaker. Studies on the evaluation of the toxicity of various salts of lead, manganese, platinum, and palladium. Environ. Health Perspect. 10:95-101, 1975.
206. Holmes, A. W. The development of the modern ammonia oxidation process: Influence of economic and technical factors in nitric acid manufacture. Platinum Met. Rev. 3:2-8, 1959.
207. Holzmann, H. Platinum recovery in ammonia oxidation plants. A new process using gold-palladium catchment gauzes. Platinum Met. Rev. 13:2-8, 1969.
208. Hoot, W. F. Production of Low-Sulfur Gasoline, EPA-650/2-74-130. Houston, Texas: M. W. Kellogg Company, 1974, 157 pp.
209. Horáček, P., and J. Drobník. Interaction of cis-dichlorodiammine-platinum(II) with DNA. Biochim. Biophys. 254:341-347, 1971.
210. Houdry, E. J. Practical catalysis and its impact on our generation. Adv. Catal. 9:499-509, 1957.
211. Houdry, J. H., and C. T. Hayes. Platinum oxidation catalysts in the control of air pollution. Platinum Met. Rev. 2:110-116, 1958.

212. Howle, J. A., and G. R. Gale. *cis*-Dichlorodiammineplatinum(II): Persistent and selective inhibition of deoxyribonucleic acid synthesis *in vivo*. Biochem. Pharmacol. 19:2757–2762, 1970.
213. Hoyt, C. D., and J. P. Ryan. Platinum-group metals, pp. 921–931. In U.S. Department of the Interior, Bureau of Mines. Minerals Yearbook, 1969. Volume I-II. Metals, Minerals, and Fuels. Washington, D.C.: U.S. Government Printing Office, 1971.
214. Hueter, F. G., G. L. Contner, K. A. Busch, and R. G. Hinners. Biological effects of atmospheres contaminated by auto exhaust. Arch. Environ. Health 12:553–560, 1966.
215. Hultgren, R., R. L. Orr, P. D. Anderson, and K. K. Kelley. Ruthenium, pp. 242–246. In Selected Values of Thermodynamic Properties of Metals and Alloys. New York: John Wiley & Sons, 1963.
216. Hume-Rothery, W. The platinum metals and their alloys: A review of their electronic structure and constitution. Platinum Met. Rev. 10:94–100, 1966.
217. Hunt, L. B. Electrical contact materials for light duty applications: Effects of oxide films on performance. Platinum Met. Rev. 1:74–81, 1957.
218. Hunter, D., R. Milton, and K. M. A. Perry. Asthma caused by the complex salts of platinum. Brit. J. Ind. Med. 2:92–98, 1945.
219. Hunter, J. B. Platinum catalysts for the control of air pollution. Platinum Met. Rev. 12:2–6, 1968.
220. Hydrogen recovery by palladium diffusion: An inexpensive large-scale process. Platinum Met. Rev. 9:50, 1965.
221. Hygienic guide series. Osmium and its compounds. Amer. Ind. Hyg. Assoc. J. 29:621–623, 1968.
222. Hysell, D. K., W. Moore, R. Hinners, M. Malanchuk, R. Miller, and J. F. Stara. Inhalation toxicology of automotive emissions as affected by an oxidation exhaust catalyst. Environ. Health Perspect. 10:57–62, 1975.
223. Hysell, D. K., W. Moore, Jr., M. Malanchuk, L. Garner, R. G. Hinners, and J. F. Stara. Comparison of Biological Effects in Laboratory Animals of Exposure to Automotive Emissions Emitted with and without Use of Catalytic Converter. Paper 74-219 Presented at 67th Annual Meeting of the Air Pollution Control Association, Denver, Colorado, June 9–13, 1974. 19 pp.
224. Hysell, D., S. Neiheisel, and D. Cmehil. Ocular irritation of two palladium and two platinum compounds in rabbits, pp. A.8.1–A.8.2. In Studies on Catalytic Components and Exhaust Emissions. Cincinnati: U.S. Environmental Protection Agency, Environmental Toxicological Research Laboratory, National Environmental Research Center, 1974.
225. Illis, A., B. J. Brandt, and A. Manson. The recovery of osmium from nickel refinery anode slimes. Metallurgical Trans. 1:431–434, 1970.
226. Improved stainless steel reactor material: Resistance to corrosion increased by platinum addition. Platinum Met. Rev. 4:149, 1960.
227. Iridium for gamma radiography: New sources of greater specific activity. Platinum Met. Rev. 6:11, 1962.
228. Ishizaka, K., and T. Ishizaka. Mechanisms of reaginic hypersensitivity: A review. Clin. Allergy 1:9–24, 1971.
229. Iwashima, K., and N. Yamagata. Environmental contamination with radioruthenium 1961–1965. J. Radiat. Res. (Tokyo) 7:91–111, 1966.
230. Johansson, S. G. O., H. Bennich, and L. Wide. A new class of immunoglobulin in human serum. Immunology 14:265–272, 1968.

References

231. Johns-Manville finds good platinum-palladium values in Montana's Stillwater. Eng. Min. J. 176(2):36, 1975.
232. Johns-Manville gets good assays from Montana platinum-palladium prospect. Eng. Min. J. 177(2):17, 1976.
233. Johnson, D. E., R. J. Prevost, J. B. Tillery, D. E. Camann, and J. M. Hosenfeld. Baseline Levels of Platinum and Palladium in Human Tissue. EPA 600/1-76-019. San Antonio, Tex.: Southwest Research Institute, 1976. 237 pp.
234. Johnston, C. Platinum mining in Alaska. Platinum Met. Rev. 6:68-74, 1962.
235. Kane-Maguire, L. A. P. The noble metals, pp. 283-375. In Inorganic Chemistry of the Transition Elements. Vol. 1. A Specialist Periodical Report. A Review of the Literature Published between October 1970 and September 1971. London: The Chemical Society, 1972.
236. Karasek, F. W. Surface analysis by ion sputtering and quadrupole mass spectrometry. Res. Dev. 24(11):40-46, 1973.
237. Karasek, F. W. Surface analysis by ISS and ESCA. Res. Dev. 24(1):25-30, 1973.
238. Karasek, S. R., and M. Karasek. The use of platinum paper, p. 97. In Report of [Illinois] Commission on Occupational Diseases to his Excellency Governor Charles S. Deneen, January, 1911. Chicago: Warner Printing Company, 1911.
239. Keil, K. Applications of the electron microprobe in geology, pp. 189-239. In C. A. Andersen, Ed. Microprobe Analysis. New York: John Wiley & Sons, 1973.
240. Kerridge, K. E. Platinum and palladium metallising preparations: Compositions and uses in the electrical and electronic industries. Platinum Met. Rev. 9:2-6, 1965.
241. Khan, A., J. M. Hill, W. Grater, E. Loeb, A. McLellan, and N. Hill. Atopic hypersensitivity to cis-dichlorodiammineplatinum(II) and other platinum complexes. Cancer Res. 35:2766-2770, 1975.
242. Kirkbright, G. F., and M. Sargent. Analytical AAS and AFS characteristics of the elements and applications data, pp. 541-718. In Atomic Absorption and Fluorescence Spectroscopy. New York: Academic Press, 1974.
243. Kistner, C. R., J. H. Hutchinson, J. R. Doyle, and J. C. Storlie. Metalolefin compounds. IV. The preparation and properties of some aryl and alkyl platinum(II)-olefin compounds, Inorg. Chem. 2:1255-1261, 1963.
244. Klimisch, R. L., and K. C. Taylor. The Catalytic Reduction of NO Over an Oxygen Treated Ruthenium Catalyst. GM Research Publication GMR-1493. Warren, Mich: General Motors Corporation Research Laboratories, 1974. 23 pp.
245. Kobylinski, T. P., B. W. Taylor, and J. E. Young. Stabilized Ruthenium Catalysts for NO_x Reduction. SAE (Soc. Automot. Eng.) Trans. 84:1089-1095, 1974.
246. Kolpakov, F. I., and A. F. Kolpakova. The effect of some metals of platinum group on the skin of experimental animals. Vestn. Dermatol. Venerol. 11:61-64, 1975. (in Russian, summary in English)
247. Kulikova, V. G. The penetration of strontium, cesium, ruthenium, and iron across placental and breast barrier. Med. Radiol. (Mosk.) 4(5):23-27, 1959. (in Russian, summary in English)
248. Kusler, D. J. Demand for Platinum to Reduce Pollution from Automobile Exhausts. Bureau of Mines Information Circular 8565. Washington, D.C.: U.S. Department of the Interior [1972]. 32 pp.
249. Kutinova, L., V. Vonka, and J. Drobnik. Inactivation of papovavirus SV 40 by cis-dichlorodiammine platinum(II). Neoplasma 19:453-458, 1972.
250. Larson, G. F., V. A. Fassel, R. H. Scott, and R. N. Kniseley. Inductively coupled plasma—optical emission analytical spectrometry. A study of some interelement effects. Anal. Chem. 47:238-243, 1975.

251. Lauder, A. Metal Oxide Catalytic Compositions. U.S. Patent 3,897,367, July 29, 1975.
252. Ledo-Dunipe, E. ¿Sensibilizacion al platino? Sensibilizacion por soldadura. Act. Dermosifiliogr. (Madrid) 48:583–584, 1957.
253. Levene, G. M., and C. D. Calnan. Platinum sensitivity: Treatment by specific hyposensitisation. Clin. Allergy 1:75–82, 1971.
254. Lever, F. M., and A. R. Powell. Ammine complexes of ruthenium. J. Chem. Soc. A1969:1477–1482.
255. Lewis, F. A. The Palladium Hydrogen System. New York: Academic Press, 1967. 178 pp.
256. Lewis, T. R., F. G. Hueter, and K. A. Busch. Irradiated automobile exhaust: Its effects on the reproduction of mice. Arch. Environ. Health 15:26–35, 1967.
257. Livingstone, S. E. Palladium complexes. Part V. Reactions of palladium compounds with 2:2' dipyridyl. J. Proc. Royal Soc. New South Wales 86:32–37, 1952.
258. Livingstone, S. E. The second- and third-row elements of Group VIII. A, B and C., pp. 1163–1370. In J. C. Bailar, Jr., H. J. Emeléus, R. Nyholm and A. F. Trotman-Dickenson, Eds. Comprehensive Inorganic Chemistry. Vol. 3. Oxford: Pergamon Press, Ltd., 1973.
259. Lupin, M. S., J. Powell, and B. L. Shaw. Transition metal-carbon bonds. Part VII. The formation of π-allylic-palladium complexes from allenes and palladium halides and the reversed reactions. J. Chem. Soc. A1966:1687–1691.
260. Lyman, T., Ed. Metals Handbook. Vol. 1. Properties and Selections of Metals. (8th ed.) Novelty, Ohio: American Society for Metals, 1961. 1236 pp.
261. MacConnell, J. D. Low temperature catalytic heaters: The cataheat range of flameless combustion systems. Platinum Met. Rev. 16:16–21, 1972.
262. Maga, J. A. Motor vehicle emissions in air pollution and their control. Adv. Environ. Sci. Technol. 2:57–89, 1971.
263. Maitlis, P. M. The Organic Chemistry of Palladium. Vol. 1. Metal Complexes. New York: Academic Press, 1971. 319 pp.
264. Maitlis, P. M. The Organic Chemistry of Palladium. Vol. 2. Catalytic Reactions. New York: Academic Press, 1971. 216 pp.
265. Malatesta, L., and M. Angoletta. Palladium(O) compounds. Part II. Compounds with triarylphosphines, triaryl phosphites, and triarylarsines. J. Chem. Soc. 1957:1186–1188.
266. Malatesta, L., and F. Bonati. Isocyanide Complexes of Metals. New York: John Wiley & Sons Ltd., 1969. 199 pp.
267. Manassen, J. Homogenous catalysis with macromolecular ligands. Platinum Met. Rev. 15:142–143, 1971.
268. Mansy, S. A. A. The Interaction of the cis- and trans-Dichlorodiammineplatinum(II) Coordination Complexes with DNA and its Components in vitro. Ph.D. Thesis. East Lansing: Michigan State University, 1972. 222 pp.
269. Mason, B. Principles of Geochemistry. (2nd ed.) New York: John Wiley & Sons, Inc., 1958. 310 pp.
270. McAuliffe, C. A., and S. G. Murray. Metal complexes of sulphur-containing amino acids. Inorgan. Chim. Acta Rev. 6:103–119, 1972.
271. Meek, S. F., G. C. Harrold, and C. P. McCord. The physiologic properties of palladium and its compounds. Ind. Med. 12:447–448, 1943.
272. Melius, P., J. E. Teggins, M. E. Friedman, and R. W. Guthrie. Inhibition of leucine aminopeptidase and malate dehydrogenase by aquoplatinum(II) complexes. Biochim. Biophys. Acta 268:194–198, 1972.

References

273. Merrin, C. A new method to prevent toxicity with high doses of cis-diammineplatinum (therapeutic efficacy in previously treated widespread and recurrent testicular tumors). Abstract C-26. In Proceedings of the AACR/ASCO (American Association for Cancer Research/American Society of Clinical Oncology) Meeting, Montreal, Canada, May, 1976.
274. Michalska, Z. M., and D. E. Webster. Supported homogeneous catalysts: Transition metal complexes with polymeric ligands. Platinum Met. Rev. 18:65–73, 1974.
275. Miller, R. G., and J. U. Doerger. Determination of platinum and palladium in biological samples. Atom. Absorpt. Newslett. 14:66–67, 1975.
276. Milne, J. E. H. A case of platinosis. Med. J. Austral. 2:1194–1195, 1970.
277. Mintern, R. A. Platinum alloy permanent magnets. Properties and applications of Platinax II. Platinum Met. Rev. 5:82–88, 1961.
278. Mintern, R. A., and J. C. Chaston. Gamma radiography with iridium[192]: Advantages in the non-destructive testing of castings and welded structures. Platinum Met. Rev. 3:12–16, 1959.
279. Mitko, F. C. Platinum-group metals, pp. 939–950. In U.S. Department of the Interior, Bureau of Mines. Minerals Yearbook, 1970. Volume I. Metals, Minerals, and Fuels. Washington, D.C.: U.S. Government Printing Office, 1972.
280. Mitko, F. C. Platinum-group metals, pp. 985–994. In U.S. Department of the Interior, Bureau of Mines. Minerals Yearbook, 1971. Volume I. Metals, Minerals, and Fuels. Washington, D.C.: U.S. Government Printing Office, 1973.
281. Monti-Bragadin, C., M. Tamaro, and E. Banfi. Mutagenic activity of platinum and ruthenium complexes. Chem. Biol. Interact. 11:469–472, 1975.
282. Moore, W., D. Hysell, L. Hall, K. Campbell, and J. Stara. Preliminary studies on the toxicity and metabolism of palladium and platinum. Environ. Health Perspect. 10:63–71, 1975.
283. Moore, W., Jr., M. Malanchuk, W. Crocker, D. Hysell, A. Cohen, and J. F. Stara. Whole body retention in rats of different ^{191}Pt compounds following inhalation exposure. Environ. Health Perspect. 12:35–39, 1975.
284. Moran, J. B. Assuring Public Health Protection as a Result of Mobile Source Emissions Control Program. SAE (Society of Automotive Engineers) Technical Paper 740285, Presented at Automotive Engineering Congress, Detroit, Michigan, Feb. 25–Mar. 1, 1974. 12 pp.
285. Movius, W. G., and R. G. Linck. Studies on the role of reduction of ruthenium(III) complexes by chromium(II) and vanadium(II). J. Amer. Chem. Soc. 92:2677–2683, 1970.
286. Müller, R. O. Trace element determination in catalysts, pp. 278–279. In Spectrochemical Analysis by X-Ray Fluorescence. (Translated by K. Keil) New York: Plenum Press, 1972.
287. Munro-Ashman, D., D. Munro, and T. H. Hughes. Contact dermatitis from palladium. Trans. St. Johns Hosp. Dermatol. Soc. 55:196–197, 1969.
288. Murray, C. Ruthenium complexes aid hydrogen process. Chem. Eng. News 53(23):17,20, 1976.
289. Nakajima, K. Odor control by UOP catalytic combustion system. Kuki Seijo (J. Jap. Air Clean. Assoc.) 9(3):13–20, 1971. (in Japanese)
290. Nast, R., and W-D. Hoerl. Komplexe acetylide von platin(II) and platin(0). Chem. Ber. 95:1478–1483, 1965.
291. National Academy of Sciences, National Academy of Engineering. Coordinating Committee on Air Quality Studies. Air Quality and Automobile Emission Control. Vol. 1. Summary Report. U.S. Senate Committee Print Serial No. 93–24. Washington, D.C.: U.S. Government Printing Office, 1974. 129 pp.

292. National Academy of Sciences, National Academy of Engineering. Coordinating Committee on Air Quality Studies. Air Quality and Automobile Emission Control. Vol. 2. Health Effects of Air Pollutants. U.S. Senate Committee Print Serial No. 93-24. Washington, D.C.: U.S. Government Printing Office, 1974. 511 pp.
293. National Academy of Sciences, National Academy of Engineering. Coordinating Committee on Air Quality Studies. Air Quality and Automobile Emission Control. Vol. 3. The Relationship of Emissions to Ambient Air Quality. U.S. Senate Committee Print Serial No. 93-24. Washington, D.C.: U.S. Government Printing Office, 1974. 137 pp.
294. National Academy of Sciences, National Academy of Engineering. Coordinating Committee on Air Quality Studies. Air Quality and Automobile Emission Control. Vol. 4. The Costs and Benefits of Automobile Emission Control. U.S. Senate Committee Print Serial No. 93-24. Washington, D.C.: U.S. Government Printing Office, 1974. 470 pp.
295. National Research Council. Commission on Sociotechnical Systems. Report by the Committee on Motor Vehicle Emissions. Washington, D.C.: National Academy of Sciences, November 1974. 190 pp.
296. National Research Council. Committee on Biologic Effects of Atmospheric Pollutants. Chromium. Washington, D.C.: National Academy of Sciences, 1974. 155 pp.
297. National Research Council. Committee on Medical and Biologic Effects of Environmental Pollutants. Nickel. Washington, D.C.: National Academy of Sciences, 1975. 277 pp.
298. National Research Council. National Materials Advisory Board. Panel on Catalysts for Automotive Emission Devices. Substitute Catalysts for Platinum in Automobile Emission Control Devices and Petroleum Refining. NMAB Report 297. Washington, D.C.: National Academy of Sciences, 1973. 94 pp.
299. Nelson, A., S. Ullberg, H. Kristoffersson, and C. Rönnbäck. Distribution of radioruthenium in mice. Acta Radiol. 58:353-360, 1962.
300. Newhouse, M. L., B. Tagg, S. J. Pocock, and A. C. McEwan. An epidemiological study of workers producing enzyme washing powders. Lancet 1:689-693, 1970.
301. New platforming catalysts: Commercial experience confirms improved stability and high yields. Platinum Met. Rev. 14:86-87, 1970.
302. Noble, M. G., and J. R. Brower. Flame Retardant Compositions. U.S. Patent 3,514,424. May 26, 1970. 5 pp.
303. Nyholm, R. S. Studies in co-ordination chemistry. Part I. Complexes of quadrivalent platinum with tertiary arsines. J. Chem. Soc. 1950:843-848.
304. Nyholm, R. S. Transition-metal complexes of some perfluoro-ligands. Chem. Soc. (Lond.) Quart. Rev. 24:1-19, 1970.
305. Nyman, C. J., C. E. Wymore, and G. Wilkinson. Reactions of tris(triphenylphosphine)platinum(O) and tetrakis(triphenylphosphine)palladium(O) with oxygen and carbon dioxide. J. Chem. Soc. A1968:561-563.
306. Oswin, H. G. Platinum metals in the fuel cell: Their function and applications in electrode structures. Platinum Met. Rev. 8:42-48, 1964.
307. Otsuka, S., Y. Tatsuno, and K. Ataka. Univalent palladium complexes. J. Amer. Chem. Soc. 93:6705-6706, 1971. (letter)
308. Padrta, F. G., P. C. Samson, J. J. Donohue, and H. Skala. Polynuclear aromatics in automobile exhaust. Amer. Chem. Soc. Div. Petrol. Chem. Prepr. 16(2):E13-E23, 1971.
309. Pallador II: A new high-output noble metal thermocouple. Platinum Met. Rev. 9:83, 1965.

References

310. Parish, W. E. A human heat-stable anaphylactic or anaphylactoid antibody which may participate in pulmonary disorders, pp. 72-90. In K. F. Austen, and L. M. Lichenstein, Eds. Asthma. Physiology, Immunopharmacology, and Treatment. New York: Academic Press, 1973.
311. Parish, W. E. Short-term anaphylactic IgG antibodies in human sera. Lancet 2:591-592, 1970.
312. Parrot, J.-L., R. Hébert, A. Saindelle, and F. Ruff. Platinum and platinosis: Allergy and histamine releatinum salts. Arch. Environ. Health 19:685-691, 1969.
313. Parrot, J.-L., A. Saindelle, and F. Ruff. Platine et platinose. Libération d'histamine par certains sels de platine et allergie au platine. Presse Med. 75:2817-2820, 1967.
314. Parrot, J.-L., A. Saindelle, and T. Tazi. Libération d'histamine par le chloroplatinate de sodium. J. Physiol. (Paris) 55:314-315, 1963.
315. Parshall, G. W., and J. J. Mrowca. σ-Alkyl and -aryl derivatives of transition metals. Adv. Organometal. Chem. 7:157-207, 1968.
316. Partington, J. R. General and Inorganic Chemistry for University Students. (3rd ed.) London: MacMillan and Co., Ltd., 1958. pp. 706-708.
317. Peavy, C. C. The importance of platinum in petroleum refining: Catalytic reforming in modern processing practice. Platinum Met. Rev. 2:48-52, 1958.
318. Penland, R. B., S. Mizushima, C. Curran, and J. B. Quagliano. Infrared absorption spectra of inorganic coordination complexes. X. Studies of some metal-urea complexes. J. Amer. Chem. Soc. 79:1575-1578, 1957.
319. Pepys, J. Atopy, pp. 877-902. In P. G. H. Gell, R. R. A. Coombs, and P. J. Lachmann, Eds. Clinical Aspects of Immunology. (3rd ed.) Oxford: Blackwell Scientific Publications, 1975.
320. Pepys, J. Hypersensitivity diseases of the lungs due to fungi and organic dusts. Monogr. Allergy 4:1-147, 1969.
321. Pepys, J., F. E. Hargreave, J. L. Longbottom, and J. Faux. Allergic reactions of the lungs to enzymes of *Bacillus subtilis*. Lancet 1:1181-1184, 1969.
322. Pepys, J., C. A. C. Pickering, and E. G. Hughes. Asthma due to inhaled chemical agents—Complex salts of platinum. Clin. Allergy 2:391-396, 1972.
323. Pepys, J., I. D. Wells, M. F. D'Souza, and M. Greenberg. Clinical and immunological responses to enzymes of *Bacillus subtilis* in factory workers and consumers. Clin. Allergy 3:143-160, 1973.
324. Petsko, G. A., D. G. Phillips, and R. J. P. Williams. The protein crystal chemistry of $PtCl_4^{-2}$. (in press)
325. Philpott, J. E. Platinum bursting discs: Applications in the protection of chemical plant. Platinum Met. Rev. 6:42-46, 1962.
326. Philpott, J. E. Surface phenomena on rhodium-platinum gauzes: Catalyst activity during ammonia oxidation. Platinum Met. Rev. 15:52-57, 1971.
327. Physical properties of the platinum metals. Platinum Met. Rev. 16:59, 1972.
328. Pickering, C. A. C. Inhalation tests with chemical allergens: Complex salts of platinum. Proc. R. Soc. Med. 65:272-274, 1972.
329. Pirie, J. M. The manufacture of hydrocyanic acid by the Andrussow process. Platinum Met. Rev. 2:7-11, 1958.
330. Pirie, J. M. The protection of chemical process equipment: The use of platinum metals for bursting discs. Platinum Met. Rev. 1:9-13, 1957.
331. Platinised titanium anodes in chlorate production. Platinum Met. Rev. 13:103, 1969.
332. Platinum and palladium in at Detroit. GM is testing combinations of the two in a 5:2 ratio, as well as ruthenium catalysts. Metals Week 43(40):1, 1972.
333. Platinum-coated titanium electrodes for cathodic protection: An electrochemical investigation. Platinum Met. Rev. 4:101, 1960.

334. Platinum gauzes for hydrogen cyanide production. Platinum Met. Rev. 11:67, 1967.
335. Platinum in hydrogen peroxide production: An improved electrolytic process. Platinum Met. Rev. 7:146, 1963.
336. Platinum-lined furnaces for plutonium production. New equipment to be installed at Windscale. Platinum Met. Rev. 5:92, 1961.
337. Platinum reforming catalysts: Production of high-octane fuels and of aromatic chemicals. Platinum Met. Rev. 5:9–12, 1961. (Summary of paper by H. Connor)
338. Pollitzer, E. L. Platinum catalysts in lead-free gasoline production: The process technology available. Platinum Met. Rev. 16:42–47, 1972.
339. Porter, R., and J. Birch, Eds. Identification of Asthma. Ciba Foundation Study Group No. 38. London: Churchill Livingstone, 1971. 179 pp.
340. Powell, A. R. The platinum metals in the periodic system: A comparative study of the transition metals. Platinum Met. Rev. 4:144–149, 1960.
341. P[reston], E. Platinum bubbler tubes in glass melting: Improvement of quality and output. Platinum Met. Rev. 7:7, 1963.
342. Preston, E. Platinum in the glass industry. Platinum Met. Rev. 4:2–9, 1960.
343. Preston, E. Platinum in the glass industry. Platinum Met. Rev. 4:48–55, 1960.
344. Preston, E. Platinum in the glass industry: The design of protective sheathing. Platinum Met. Rev. 10:78–83, 1966.
345. Price, R. The platinum resistance thermometer: A review of its construction and applications. Platinum Met. Rev. 3:78–87, 1959.
346. Priddis, J. E. The design of platinum-wound electric resistance furnaces. Platinum Met. Rev. 2:38–44, 1958.
347. Process survey. Nitric acid. The technology,costs and performance of today's chief manufacturing methods available for licensing. Eur. Chem. News 17(417) (Supplement) Jan. 30, 1970. 50 pp.
348. Production of ultra-pure hydrogen: A dissociated ammonia diffusion plant. Platinum Met. Rev. 8:91, 1964.
349. Public Law 91–604. An act to amend the Clean Air Act to provide for a more effective program to improve the quality of the nation's air. U.S. Statutes at Large 84:1676–1713, 1971.
350. Public Law 93–319. An act to provide for means of dealing with energy shortages by requiring reports with respect to energy resources, by providing for temporary suspension of certain air pollution requirements, by providing for coal conversion, and for other purposes. U.S. Statutes at Large 88:246–265, 1974.
351. Quinn, T. J., and T. R. D. Chandler. Platinum metal thermocouples: New international reference tables. Platinum Met. Rev. 16:2–9, 1972.
352. Rankama, K., and Th. G. Sahama. Geochemistry. Chicago: The University of Chicago Press, 1950. 912 pp.
353. Raub, E. Metals and alloys of the platinum group. J. Less Common Metals 1:3–18, 1959.
354. Reducing infra-red radiation in mercury vapour lamps. Platinum Met. Rev. 2:128, 1958.
355. Reinhardt, R. A., and W. W. Monk. The kinetics of the successive ammonation reactions of tetrachloropalladate(II) ion. Inorg. Chem. 9: 2026–2030, 1970.
356. Renshaw, E., and A. V. Thomson. Trace studies to locate the site of platinum ions within filamentous and inhibited cells of *Escherichia coli*. J. Bacteriol. 94:1915–1918, 1967.
357. Reslova, S. The induction of lysogenic strains of *Eschericia coli* by *cis*-dichlorodiammineplatinum(II). Chem. Biol. Interact. 4:66–70, 1971/1972.

References

358. Rhodes, D., P. W. Piper, and B. F. C. Clark. Location of a platinum binding site in the structure of yeast phenylalanine transfer RNA. J. Mol. Biol. 89:469–475, 1974.
359. Rhodium plated uniselectors: Improved contact performance at low voltages. Platinum Met. Rev. 4:65, 1960.
360. Rhodium plating in lighthouse beacons. Platinum Met. Rev. 7:24, 1963.
361. Richards, A. E. New type platforming catalyst produced and tested in Europe. Platinum Met. Rev. 2:23–27, 1958.
362. Riddle, J. L., G. T. Furukawa, and H. H. Plumb. Platinum Resistance Thermometry. National Bureau of Standards Monograph 126. Washington, D.C.: U.S. Government Printing Office, 1973. 126 pp.
363. Ridgway, L. P., and D. A. Karnofsky. The effects of metals on the chick embryo: Toxicity and production of abnormalities in development. Ann. N. Y. Acad. Sci. 55:203–215, 1952.
364. Roberts, A. E. Platinosis: A five year study of the effects of soluble platinum salts on employees in a platinum laboratory and refinery. A.M.A. Arch. Ind. Hyg. Occup. Med. 4:549–559, 1951.
365. Roberts, J. J. Bacterial, viral and tissue culture studies on neutral platinum complexes. Recent Results Cancer Res. 48:79–97, 1974.
366. Roberts, J. J., and J. M. Pascoe. Cross-linking of complementary strands of DNA in mammalian cells by antitumour platinum compounds. Nature 235:282–284, 1972.
367. Roberts, P. M., and D. A. Stiles. Palladium alloy diffusion units: A new range of commercial equipment for the production of ultra-pure hydrogen. Platinum Met. Rev. 13:141–145, 1969.
368. Robertson, A. J. B. The early history of catalysis. Platinum Metals Rev. 19:64–69, 1975.
369. Robertus, J. D., J. E. Ladner, J. T. Finch, D. Rhodes, R. S. Brown, B. F. C. Clark, and A. Klug. Structure of yeast phenylalanine tRNA at 3 Å resolution. Nature 250:546–551, 1974.
370. Robins, A. B. Interactions with biomacromolecules. Recent Results Cancer Res. 48:63–78, 1974.
371. Robins, A. B. The reaction of ^{14}C-labelled platinum ethylenediamine dichloride with nucleic acid constituents. Chem. Biol. Interact. 6:35–45, 1973.
372. Role of metal complexes and metal salts in cancer chemotherapy. A symposium presented at the 30th Southwest regional meeting of the American Chemical Society, Houston, Texas, December 10–11, 1974. Cancer Chemother. Rep. 59:587–673, 1975.
373. Rosenberg, B. Platinum coordination complexes in cancer chemotherapy. Naturwissenschaften 60:399–406, 1973.
374. Rosenberg, B., E. Renshaw, L. van Camp, J. Hartwick, and J. Drobnik. Platinum-induced filamentous growth in *Escherichia coli*. J. Bacteriol. 93:716–721, 1967.
375. Rosenberg, B., L. van Camp, E. B. Grimley, and A. J. Thomson. The inhibition of growth on cell division in *Escherichia coli* by different ionic species of platinum(IV) complexes. J. Biol. Chem. 242:1347–1352, 1967.
376. Rosenberg, B., L. van Camp, and T. Krigas. Inhibition of cell division in *Escherichia coli* by electrolysis products from a platinum electrode. Nature 205:698–699, 1965.
377. Rosenberg, B., L. van Camp, J. E. Trosko, and V. H Mansour. Platinum compounds: A new class of potent antitumour agents. Nature 222: 385–386, 1969.
378. Roth, J. F. The production of acetic acid: Rhodium catalysed carbonylation of methanol. Platinum Met. Rev. 19:12–14, 1975.

379. Roth, J. F., and R. C. Doerr. Oxidation-reduction catalysis. Ind. Eng. Chem. 53:293–296, 1961.
380. Roydhouse, R. H. Materials in Dentistry. A Discussion for the Users of Dental Materials. Chicago: Year Book Publishers, Inc., 1962. 210 pp.
381. Rylander, P. N. Catalytic Hydrogenation Over Platinum Metals. New York: Academic Press, 1967. 550 pp.
382. Rylander, P. N. Dehydrogenation, pp. 1–59. In Organic Syntheses with Noble Metal Catalysts. New York: Academic Press, 1973.
383. Rylander, P. N. Homogeneous hydrogenation, pp. 60–76. In Organic Syntheses with Noble Metal Catalysts. New York: Academic Press, 1973.
384. Rylander, P. N. Osmium and ruthenium tetroxides as oxidation catalysts, pp. 121–144. In Organic Syntheses with Noble Metal Catalysts. New York: Academic Press, 1973.
385. Rylander, P. N. Palladium catalysts in organic chemistry, pp. 159–181. In E. M. Wise. Palladium. Recovery, Properties and Uses. New York: Academic Press, 1968.
386. S., A. J. Electrical contact phenomena: The fifth international conference. Platinum Met. Rev. 14:103–104, 1970.
387. S., L. L. Surface treatment of titanium with palladium: An economical corrosion prevention process. Platinum Met. Rev. 14:47, 1970.
388. Sagert, N. H., and R. M. L. Pouteau. The production of heavy water: Hydrogen-water deuterium exchange over platinum metals on carbon supports. Platinum Met. Rev. 19:16–21, 1975.
389. Saindelle, A., and F. Ruff. Histamine release by sodium chloroplatinate. Brit. J. Pharmacol. 35:313–321, 1969.
390. Sakakibara, M., Y. Takahashi, S. Sakai, and Y Ishii. Preparation of π-allylic palladium complexes from tin(II)chloride, allylic halides, and palladium salts. J. Chem. Soc. (Lond.) D1969:396–397.
391. Sandell, E. B. Colorimetric Determination of Traces of Metals. (3rd ed.) New York: Interscience Publishers, Inc., 1959. 1,032 pp.
392. Schlatter, J. C., R. L. Klimisch, and K. C. Taylor. Exhaust catalysts: Appropriate conditions for comparing platinum and base metals. Science 179:798–799, 1973.
393. Schmidt, L. D., and D. Luss. Physical and chemical characterization of platinum-rhodium gauze catalysts. J. Catalysis 22:269–279, 1971.
394. Schroeder, H. A., and M. Mitchener. Scandium, chromium(VI), gallium, yttrium, rhodium, palladium, indium in mice: Effects on growth and life span. J. Nutr. 101:1431–1437, 1971.
395. Schroeder, H. A., and A. P. Nason. Interactions of trace metals in mouse and rat tissues; zinc, chromium, copper and manganese with 13 other elements. J. Nutr. 106:198–203, 1976.
396. Schwartz, L., L. Tulipan, and S Peck. Occupational Diseases of the Skin. (2nd ed.) Philadelphia: Lea and Febiger, 1947. 964 pp.
397. Scott, K. G., and J. Crowley. Tracer studies, pp. 8–18. In Medical and Health Physics Quarterly Report, University of California, [Berkeley], Radiation Laboratory, April, May, June, 1951.
398. Searles, R. A. Pollution from nitric acid plants: Purification of tail gas using platinum catalysts. Platinum Met. Rev. 17:57–63, 1973.
399. Selman, G. L., J. G. Day, and A. A. Bourne. Dispersion strengthened platinum: Properties and characteristics of a new high temperature material. Platinum Met. Rev. 18:46–57, 1974.

400. Sercombe, E. J. Exhaust purifiers for compression ignition engines: Catalytic control of diesel exhaust gases. Platinum Met. Rev. 19:2–11, 1975.
401. Sergi, S., V. Marsale, R. Pietropaolo, and F. Faraone. Some new σ-bonded arylplatinum complexes. J. Organometal. Chem. 23:281–284, 1970.
402. Shaw, B. L. A revised structure for butadienepalladous chloride. Chem. Ind. (Lond.) 1962:1190.
403. Shelef, M., and H. S. Gandhi. Ammonia formation in catalytic reduction of nitric oxide by molecular hydrogen. I. Base metal oxide catalysts. Ind. Eng. (Prod. Res. Develop.) 11:2–11, 1972.
404. Shelef, M., and H. S. Gandhi. The reduction of nitric oxide in automobile emissions: Stabilization of catalysts containing ruthenium. Platinum Met. Rev. 18:2–14, 1974.
405. Shimazu, M., and B. Rosenberg. A similar action to UV-irridation and a preferential inhibition of DNA synthesis in *E. coli* by antitumor platinum compounds. J. Antibiot. 26:243–245, 1973.
406. Shingledecker, R. A. One Component Non-toxic Self-extinguishing Silicone Elastomer. U.S. Patent 3,734,881. May 22, 1973. 6 pp.
407. Shoobert, G. W. Iridium electrodes increase spark plug life: Resistance to attack by lead compounds. Platinum Met. Rev. 6:92–94, 1962.
408. Shooter, K. V., R. Howse, R. K. Merrifield, and A. B. Robins. The interaction of platinum II compounds with bacteriophages T7 and R17. Chem. Biol. Interact. 5:289–307, 1972.
409. Shulman, A., and F. P. Dwyer. Metal chelates in biological systems, pp. 383–439. In F. P. Dwyer and D. P. Mellor, Eds. Chelating Agents and Metal Chelates. New York: Academic Press, 1964.
410. Sloboda, M. H. High purity palladium brazing alloys: Multi-stage jointing in the manufacture of thermionic valves. Platinum Met. Rev. 7:8–11, 1963.
411. S[loboda], M. H. Platinum alloys for brazing tungsten: Fabrication for high temperature service. Platinum Met. Rev. 7:56–57, 1963.
412. Smidt, J. Oxidation of olefins with palladium chloride catalysts. J. Chem. Ind. (Lond.) 1962:54–61.
413. Smidt, J., W. Hafner, R. Jira, J. Sedlmeier, R. Sieber, R. Rüttinger, and H. Kojer. Katalytische Umsetzungen von Olefinen an Platinmetall-Verbindungen: Das Consortium-Verfahren zur Herstellung von Acetaldehyd. Angew. Chem. 71:176–182, 1959.
414. Smidt, J., W. Hafner, R. Jira, R. Sieber, J. Sedlmeier, and A. Sabel. Olefinoxydation mit Palladiumchlorid-Katalysatoren. Angew. Chem. 74:93–102, 1962.
415. S[mith], F. J. Palladium addition protects titanium in hot concentrated chloride solutions. Platinum Met. Rev. 12:53, 1968.
416. S[mith], F. J. Palladium-titanium alloy in chemical plant. Platinum Met. Rev. 13:67, 1969.
417. Smith, F. J. Standard kilogram weights: A story of precision fabrication. Platinum Met. Rev. 17:66–68, 1973.
418. Smith, I. C., B. L. Carson, and T. L. Ferguson. Osmium: An appraisal of environmental exposure. Environ. Health Perspect. 8:201–213, 1974.
419. Smith, I. C., and T. L. Ferguson. Osmium. An Appraisal of Environmental Exposure. (Prepared for the National Institute of Environmental Health Sciences) Kansas City, Missouri: Midwest Research Institute, 1973. 50 pp.
420. Snell, A. K. The use of platinum in high power thermionic valves. Platinum Met. Rev. 4:82–85, 1960.

421. Speer, R. J., H. Ridgway, L. M Hall, D. P. Steward, K. E. Howe, D. Z. Lieberman, A. D. Newman, and J. M. Hill. Coordination complexes of platinum as antitumor agents. Cancer Chemother. Rep. I 59:629–641, 1975.
422. Spikes, J. D., and C. F. Hodgson. Enzyme inhibition by palladium chloride. Biochem. Biophys. Res. Com. 35:420–422, 1969.
423. Spode, E., and F. Gensicke. Zur Frage des Stoffwechsels von Radioruthenium in der weissen Maus. I. Verteilung und Ausscheidung von trägerfreiem ^{106}Ru bei unterschiedlicher Applikationsart. Strahlentherapie 111:266–272, 1960.
424. Squire, J. R. Tissue reactions to protein sensitization. Brit. Med. J. 1:1–7, 1952
425. Stara, J. F., N. S. Nelson, H. L. Krieger, and B. Kahn. Gastrointentinal absorption and tissue retention of radioruthenium, pp. 307–318. In S. C. Skoryna and D. Waldron-Edward, Eds. Intestinal Absorption of Metal Ions, Trace Elements, and Radionuclides. New York: Pergamon Press, 1970.
426. Stara, J. F., N. S. Nelson, R. J. D. Rosa, and L. K. Bustad. Comparative metabolism of radionuclides in mammals: A review. Health Phys. 20:113–137, 1971.
427. Stenius, B., L. Wide, W. M. Seymour, V. Holford-Strevens, and J. Pepys. Clinical significance of specific IgE to common allergens: I. Relationship of specific IgE against *Dermatophagoides* ssp. and grass pollen to skin and nasal tests and history. Clin. Allergy 1:37–55, 1971.
428. Sterba, M. J. The Impact of Catalytic Processing on Motor Fuel Quality Trends. Paper 10C Presented at the 68th National Meeting of the American Institute of Chemical Engineers, Houston, Texas, Feb. 28-March 4, 1971. 18 pp.
429. S[tevenson], J. A. Platinum-lined furnace for the fluorination of uranium compounds. Features of design and construction. Platinum Met. Rev. 8:12–13, 1964.
430. Stiles, D. A., and P. H. Wells. The production of ultra-pure hydrogen: Quality control of palladium alloy diffusion units. Platinum Met. Rev. 16:124–128, 1972.
431. Stone, P. J., A. D. Kelman, and F. M. Sinex. Specific binding of antitumour drug cis-Pt(NH$_3$)$_2$Cl$_2$ to DNA rich in guanine and cytosine. Nature 251:736–737, 1974.
432. Stuart, B. O., and J. C. Gaven. Ruthenium oxide inhalation studies, pp. 59–62. In R. C. Thompson and E. G. Swezea, Eds. Pacific Northwest Laboratory Annual Report for 1965 in the Biological Sciences. BNWL–280. Richland, Wash.: Battelle Pacific Northwest Laboratory, 1966.
433. Stupfel, M., M. Magnier, F. Romary, M-H. Tran, and J-P. Moutet. Lifelong exposure of SPF rats to automotive exhaust gas. Arch. Environ. Health 26:264–269, 1973.
434. T., G. Monolithic ceramic capacitors: Platinum metal electrodes in a fired multilayer construction. Platinum Met. Rev. 12:46–47, 1968.
435. Taylor, R. T., and M. L. Hanna. Methylcobalamin: Methylation of platinum and demethylation with lead. J. Environ. Sci. Health A11:201–211, 1976.
436. The new post office relay. Platinum Met. Rev. 15:141, 1971.
437. Thompson, R. C., M. H. Weeks. O. L. Hollis, J. E. Ballou, and W. D. Oakley, Metabolism of radio-ruthenium in the rat. Consideration of permissible exposure limits. Amer. J. Roentgenol. Radium Ther. Nucl. Med. 79:1026–1044, 1958.
438. Thomson, A. J. The interactions of platinum compounds with biological molecules. Recent Results Cancer Res. 48:38–62, 1974.
439. Thomson, A. J., R. J. P. Williams, and S. Reslova. The chemistry of complexes related to cis-Pt(NH$_3$)$_2$Cl$_2$ an anti-tumor drug. Struct. Bond. 11:1–46, 1972.
440. Thornton, D. P., Jr. Platinum oxidation catalysts in the control of air pollution. Platinum Met. Rev. 7:82–87, 1963.

441. Tillery, J. B., and D. E. Johnson. Determination of platinum, palladium, and lead in biological samples by atomic absorption spectrophotometry. Environ. Health Perspect. 12:19-26, 1975.
442. Tong, S. S. C., R. A. Morse, C. A. Bache, and D. J. Lisk. Elemental analysis of honey as an indicator of pollution. Arch. Environ. Health 30:329-332, 1975.
443. Traynor, J. E., and S. W. Leeper. Metabolism of Ruthenium in the Rat. Technical Documentary Report AFSWC-TDR-61-105. Kirtland Air Force Base, N. Mex.: U.S. Air Force Special Weapons Center, 1961. 13 pp.
444. Tschugajeff, L. Über Pentaminverbindungen des vierwertigen Platins. Z. Anorgan. Allg. Chem. 137:1-31, 1924.
445. Tsuji, J., and K. Ohno. Decarbonylation reactions using transition metal compounds. Synthesis 1969:157-169.
446. Tugwell, G. L. Industrial applications for the noble metals. Metal Progr. 88(1):73-78, 1965.
447. Tugwell, G. L. Uses for noble metals in industry. Metal Progr. 87(6):79-84, 1965.
448. Uhlig, H. H., Ed. The Corrosion Handbook. New York: John Wiley & Sons, Inc., 1948. 1188 pp.
449. Ultra-pure hydrogen from water: An electrolytic diffusion cell. Platinum Met. Rev. 12:15, 1968.
450. U.S. Bureau of Mines. 1970 Edition. Minerals Facts and Problems. Bureau of Mines Bulletin 650. Washington, D.C.: U.S. Department of the Interior, 1970. 1,291 pp.
451. U.S. Department of Health, Education, and Welfare. Control of air pollution from new motor vehicles and new motor vehicle engines. Federal Register 35:17288-17313, 1970.
452. U.S. Department of Labor. Occupational Safety and Health Administration. Subpart G. Occupational Health and Environmental Control. Air contaminants. Federal Register 37:22139-22144, 1972.
453. U.S. Environmental Protection Agency. Chrysler Corp., Ford Motor Co., and General Motors Corp. Applications for suspension of 1977 motor vehicle emission standards; decision of the administrator. Federal Register 40:11900-11916, 1975.
454. U.S. Environmental Protection Agency. Control of air pollution from new motor vehicles and new motor vehicle engines. Federal Register 36:12652-12664, 1971.
455. U.S. Environmental Protection Agency. Issue Paper. Estimated Changes in Human Exposure to Suspended Sulfate Attributable to Equipping Light Duty Motor Vehicles with Oxidation Catalysts. Effects of Particulate Sulfates on Human Health. Automotive Sulfate Emissions. Jan. 11, 1974. 54 pp. (see Federal Register 39:9229-9231, 1974)
456. Vallery-Radot, P., and P. Blamoutier. Sensibilisation au chloroplatinite de potassium. Accidents graves de choc survenus à la suite d'une cutiréaction avec ce sel. Bull. Mem. Soc. Med. Hopit. 3 série. 53:222-230, 1929.
457. Vaska, L. Reversible activation of covalent molecules by transition metal complexes. The role of the covalent molecule. Accounts Chem. Res. 1:335-344, 1968.
458. Venugopal, B., and T. P. Luckey. Toxicology of nonradioactive heavy metals and their salts, pp. 4-73. In T. D. Luckey, B. Venugopal, and D. Hutcheson. Heavy Metal Toxicity, Safety and Hormology. In F. Coulston and F. Korte. Environmental Quality and Safety. Supplement Vol. 1. New York: Academic Press, 1975.
459. Vines, R. F., and E. M. Wise, (Ed.) The Platinum Metals and Their Alloys. New York: The International Nickel Company, Inc., 1941. 141 pp.
460. Voorhoeve, R. J. H. The detection of undesirable constituents from the reduction of NO over platinum catalysts, Abstract COLL 37. In Abstracts of Papers. 170th

National Meeting. American Chemical Society, Chicago, Illinois, August 24–29, 1975.
461. Wallace, R. Electro-matic road traffic control equipment: Development of the pneumatic contactor. Platinum Met. Rev. 2:12–15, 1958.
462. Walsh, T. J., and E. A. Hausman. The platinum metals, pp. 379–511. In I. M. Kolthoff, P. J. Elving, and E. B. Sandell, Eds. Treatise on Analytical Chemistry. Part II. Analytical Chemistry of the Elements. Vol. 8. The Rare Earths. Bi, V, Cr, The Platinum Metals. New York: Interscience Publishers, 1963.
463. Want, J. G. The design of light duty electrical contacts: Economics of manufacturing and assembly methods. Platinum Met. Rev. 5:42–50, 1961.
464. Waters, M. D., T. O. Vaughan, D. J. Abernethy, H. R. Garland, C. C. Cox, and D. L. Coffin. Toxicity of platinum (IV) salts for cells of pulmonary origin. Environ. Health Perspect. 12:45–56, 1975.
465. Watt, G. W., M. T. Walling, Jr., and P. I. Mayfield. Evidence for the existence of an ammine of platinum(O). J. Amer. Chem. Soc. 75:6175–6177, 1953.
466. Webb, G. Ruthenium and osmium as hydrogenation catalysts. Platinum Met. Rev. 8:60–66, 1964.
467. Webber, C. E., and J. W. Harvey. Accidental human inhalation of ruthenium tetroxide. Health Phys. 30:352–355, 1976.
468. Webster, B. Honeybees aiding pollution fight. New York Times 125:41, Sept. 24, 1975.
469. Wei, J. Catalysis for motor vehicle emissions. Adv. Catal. 24:57–129, 1975.
470. Wei, J. How thick should the catalytic layer be? Abstract INDE 109. In Abstracts of Papers. 167th National Meeting, American Chemical Society, Los Angeles, California, March 31-April 5, 1974.
471. Wells, P. B. The platinum metals as selective hydrogenation catalysts: A basic approach. Platinum Met. Rev. 7:18–23, 1963.
472. West, J. M. Platinum-group metals, pp. 1043–1054. In U.S. Department of the Interior, Bureau of Mines. Minerals Yearbook, 1972. Metals, Minerals, and Fuels. Washington, D.C.: U.S. Government Printing Office, 1974.
473. Wester, P. O. Concentration of 24 trace elements in human heart tissue determined by neutron activation analysis. Scand. J. Clin. Lab. Invest. 17:357–370, 1965.
474. Westland, A. D., and M. Northcott. Aryl complexes of platinum(II). A study of the bonding between platinum and tertiary phosphine, arsine, and stibine. Can J. Chem. 48:2907–2910, 1970.
475. Wichers, E., W. G. Schlecht, and C. L. Gordon. Attack of refractory platiniferous materials by acid mixtures at elevated temperatures. J. Res. Nat. Bur. Stand. 33:363–381, 1944.
476. Wide, L., H. Bennich, and S. G. O. Johansson. Diagnosis of allergy by an in-vitro test for allergen antibodies. Lancet 2:1105–1107, 1967.
477. Wide, L., and L. Juhlin. Detection of penicillin allergy of the immediate type by radioimmunoassay of reagins (IgE) to penicilloyl conjugates. Clin. Allergy 1:171–177, 1971.
478. Wiester, M. J. Cardiovascular actions of palladium compounds in the unanesthetized rat. Environ. Health Perspect. 12:41–44, 1975.
479. Wilke, G., B. Bogdanović, P. Hardt, P. Heimbach, W. Keim, M. Kröner, W. Oberkirch, K. Tanaka, E. Steinrücke, D. Walter and H. Zimmermann. Ally-transition metal systems. Angew. Chem. Int. Ed. 5:151–164, 1966.
480. Wilkinson, R. G. Removal of chloride contaminants from nitric acid: Electrolytic process uses platinum anodes. Platinum Met. Rev. 5:128–131, 1961.

References

481. Williams, D. R. Anticancer drug design involving complexes of amino-acids and metal ions. Inorgan. Chim. Acta Rev. 6:123–131, 1972.
482. Willis, K. J. The design of precision wire-wound potentiometers. Platinum Met. Rev. 2:74–82, 1958.
483. Wilson, B. J., Ed. The Radiochemical Manual (2nd ed.) Amersham, England: The Radiochemical Centre, 1966. 327 pp.
484. Wiltshaw, E., and T. Kroner. Phase II study of cis-dichlorodiammineplatinum(II) (NSC-119875) in advanced adenocarcinoma of the ovary. Cancer Treat. Rep. 60:55–60, 1976.
485. Wise, E. M. Palladium. Recovery, Properties and Uses. New York: Academic Press, 1968. 187 pp.
486. Wittes, R. E., E. Cvitkovic, I. H. Krakoff, and E. W. Strong. DDP in epidermoid carcinoma. The role of cis-diamminedichloroplatinum(II) in the treatment of head and neck cancer. J. Clin. Hematol. Oncol. (Wadley Institutes of Molecular Medicine) 7:711–716, 1977.
487. Wood, J. M. Biological cycles for toxic elements in the environment. Science 183:1049–1052, 1974.
488. Woodman, R. J., A. E. Sirica, M. Gang, I. Kline, and J. M. Venditti. The enhanced therapeutic effect of cis-platinum(II) diamminodichloride against L1210 leukemia when combined with cyclophosphamide or 1,2-bis-(3,5-dioxopiperazine-1-yl) propane or several other antitumor agents. Chemotherapy 18:169–183, 1973.
489. Woron, W. Catalyst car sets wild oats afire. Automot. News July 21, 1975:36.
490. Wright, T. L., and M. Fleischer. Geochemistry of the Platinum Metals. Geological Survey Bulletin 1214-A. Washington, D.C.: U.S. Government Printing Office, 1965. 24 pp.
491. Wynn, N. Platinum in the decoration of ceramic wares. Platinum Met. Rev. 3:60–65, 1959.
492. Yagoda, A., R. C. Watson, J. C. Gonzalez-Vitale, H. Grabstald, and W. F. Whitmore. Cis-dichlorodiammineplatinum(II) in advanced bladder cancer. Cancer Treat. Rep. 60:917–923, 1976.
493. Yamagata, N., K. Iwashima, T. A. Iinuma, K. Watari, and T. Nagai. Uptake and retention experiments of radioruthenium in man—I. Health Phys. 16:159–166, 1969.

Index

Absorption of platinum metals, 173
Acids, reaction with platinum metals, 46–49, 67–68. *See also* Nucleic acid
Air. *See* Ambient air
Albuminuria, from palladium chloride, 81
Alkalis, reaction with platinum metals, 68
Allergy to platinum metals
 asthmatic, 100, 102, 121
 autoallergic, 117
 causes of, 174
 delayed tuberculin-type, 121
 dermatologic, 102, 103
 immediate, 106
  IgE, 107
  radioallergosorbent tests for, 115–16
  skin testing for, 107–14
 immune-complex complement-dependent, 117–19
  bronchial tests for, 119–21
  nasal tests for, 119
 in catalyst manufacturing plants, 164
 in petroleum refineries, 9, 186
 in platinum refineries, 109, 114, 115, 133, 172
 suggested testing for, 177, 178

Alloys, platinum-metal, 3, 4
 brazing of, 37–38
 in bushings, 28–29
 in furnaces, 26
 in strain gauges, 27
 properties of, 42
Ambient air, concentration of platinum metals in, 126, 127, 167, 178
Ammonia
 emitted from catalytic converters, 138, 160
 oxidation of, 19
Amperometry, for determining platinum metals, 77
Animals
 effect of platinum drugs on tumors in, 31, 32, 35
 experiments on platinum toxicity in, 80, 81, 101–2
 platinum-metal concentration in, 134
 retention of platinum metals by, 95–99
Aqua regia, for dissolving platinum metals, 67–68
Aromatics, from catalytic reforming, 196–97
Atomic-absorption spectroscopy, to de-

termine platinum metals, 71–72, 166
"Atopy" reactivity, 107–11, 123, 169
Automobiles. *See also* Catalytic converters
  exhaust emission from, 137
  platinum metals in voltage regulators of, 23
Autopsy samples, to determine platinum-metal concentration, 129, 132, 169

Bacterial and viral effects of platinum-metal complexes, 89
  bactericidal effects, 90–91
  filamentation effects, 91–93
  induction of lysogenic bacteria, 93
  viral inactivation, 94–95
Bivalent metal complexes, 58–61
Blood
  effect of platinum metals on, 81
  platinum concentration in, 128–29, 133, 169
Bone marrow, effect of platinum metals on, 80, 81
Braggite, 3
Bushings, platinum alloys used in, 28–29

Canada, platinum metals in, 5, 6
Cancer, platinum metals to arrest, 2, 29–35, 79, 165, 167, 174–75
Capacitors, platinum metals used in, 23–24
Catalysts. *See also* Catalytic converters; Catalytic reforming
  bimetallic, 18, 184
  in petroleum refining, 181, 184
  platinum metals as, 1, 14–15, 18–23, 166
  problems in sampling, 66–67
  uses of, 164
Catalytic converters, 2, 18–19, 165, 175
  beneficial effects of, 148–51, 170–71
  demands placed on, 137–38
  effect on environment, 79, 135
  forms of, 142–43
  geometry of, 142
  legislation for, 135–36
  oxidation catalyst for, 140–46
  problems from, 138, 171
    ammonia emission, 138, 160

    emission of particulate matter, 157–58
    fire hazard, 160, 162
    hydrogen cyanide emission, 159
    overheating, 160, 161, 172
    platinum requirements, 162–63
    sulfuric acid emission, 151–57, 171
  reduction catalysts for, 146–47
  suggested improvements in, 139–40
  three-way catalysts for, 147–48
Catalytic reforming, 184, 186–87
  aromatics produced by, 196–97
  byproducts from, 197
  capacity, 192–94
  contributions to U.S. gasoline pool, 195–96
  description of process, 190–92
  reactions in, 188–89
  regeneration in, 191–92
  worldwide, 192–94
Ceramic honeycomb material, for exhaust-gas purification, 18–19
Chemical analysis of platinum metals
  methods for, 71–78
  preparation of samples for, 66–68
  separation of metals for, 69–70
Chemical industry, platinum metals as catalysts in, 1, 164
Chemical reaction of platinum metals, 44–45. *See also* Compounds of platinum metals
  with acids, 46, 48–49
  with halogens, 46
  with hydrogen, 46
  with oxygen, 45–46
Chemotherapy, platinum metals in, 2, 29, 79, 89, 174–75
Chlorine, reaction with platinum metals, 47, 68
Chloroplatinic acid, 49
Chromatography, for separating platinum metals, 70
Cluster compounds, 65
Coal, platinum metals in, 126, 128
Complexes
  bivalent metal, 58–61
  iridium, 64–65
  organometallic, 62–63
  platinum
    bactericidal effects of, 89–91, 168
    filamentation effects of, 91–93

Index

induction of lysogenic bacteria by, 93
interaction with amino acids, 85–86
interaction with enzymes, 84–85
interaction with nucleic acid, 86–89, 168
interaction with proteins, 82–83, 168
viral inactivation by, 94–95
quadrivalent metal, 61–62
rhodium, 64–65
ruthenium, 63–64
zerovalent metal, 58, 59
Compounds of platinum metals
binary, 49–57
coordination, 57–63, 167
multimetallic cluster, 65
Concentration of platinum metals, 169
autopsy data on, 129, 132, 169
difficulty in obtaining baseline data on, 126, 170, 177–78
in animals, 134
in earth's crust, 1, 3, 4, 125
in man, 128–29, 131–33
in rocks, 4–5
in southern California, 130
in vegetation, 133–34
Cooperite, 3
Corrosion, platinum-metal resistance to, 14, 27–28
Coulombmetry, for determining platinum metals, 77

Dental uses of platinum metals, 15, 35
Determination of platinum metals, 166
electrochemical methods for
amperometry, 77
coulombmetry, 77
polarography, 76
gravimetric methods for, 77
neutron-activation analysis for, 75–76
spectrochemical methods for
atomic-absorption spectroscopy, 71–72
electron-probe microanalysis, 75
emission spectroscopy, 71
inductively coupled plasma spectroscopy, 73
spark-source mass spectrometry, 74–75
spectrophotometry, 73
surface analysis, 75
x-ray fluorescence spectroscopy, 74

volumetric methods for, 77–78
cis-Dichlorodiammineplatinum
allergic reaction to, 115
antitumor effects of, 30–35, 57, 87, 168
DNA and, 87–88
effect on lysogenic bacteria, 93
filamentation effect of, 91–92
preliminary clinical trials of, 36
viral effects of, 94–95
Diuresis, from palladium chloride, 81
Divalent platinum, 80
DNA, reactions with platinum metals, 86–89, 92, 168

Earth's crust, platinum metals in, 1, 3, 4, 125
Electric industry, platinum metals in, 1, 23–25, 166–67
Electrochemical methods for determining platinum metals, 76–77
Electromechanical electrodes, platinum used in, 24
Electron-probe microanalysis, to determine platinum metals, 3, 75
Emission
of platinum metals
from catalytic converters, 2, 134*ff*
from smelters, 163
of sulfuric acid, 2, 151–57
Emission spectroscopy, for determining platinum metals, 71, 166
Enzymes, inhibition by platinum complexes, 84–85
Exhaust-gas control, catalytic application of platinum in, 18–19
Exposure to platinum metals
chronic, 101–4
environmental, 106, 169
from inhalation, 100–101, 168
Eyes, effect of palladium compounds on, 81–82, 102

Feces, platinum concentration in, 128, 133
Filamentation effects of platinum-metal complexes, 91–93
Fire assay, for separating platinum metals, 69
Flame retardants, platinum compounds used in, 27
Fluorine, reaction of platinum metals with, 47, 51, 54

Fuel cells, platinum used in, 25
Furnaces, platinum alloys used in, 26

Gasoline. *See also* Refining of petroleum
  classes of, 195
  leaded, 178
  unleaded, 149, 181
  U.S. pool of, 195-96
Gauges, strain, platinum alloys used in, 27
Glass industry, use of platinum metals in, 15, 29
Gold
  alloyed with platinum metals, 35
  recovery of, in nickel-refining process, 11
Gravimetric methods for determining platinum metals, 77

Hair, platinum concentration in, 128, 133
Halogens, reaction with platinum metals, 46, 47
Health effects of platinum metals, 1, 172. *See also* Allergy to platinum metals
Heat, effect on platinum metals, 45-46
Hemolysis, from palladium chloride, 81
Histamine liberation, by platinum salts, 116-17, 174
Hydrogen
  palladium to separate other gases from, 35, 37
  platinum used in production of, 25
  reaction of platinum metals with, 46
Hydrogenation, platinum metals as catalysts in, 20-21
Hydrogen cyanide
  from catalytic converters, 159
  use of platinum in manufacture of, 20

IgE allergy, 107, 115-16
Inductively coupled plasma spectroscopy, for determining platinum metals, 73
Industrial uses of platinum metals, 1, 14-15, 18, 22-23, 27-29, 37-38, 164, 166. *See also* Uses of platinum metals
Inhalation of platinum metals
  bronchial and nasal tests for, 119-20
  from catalytic converters, 158
  most frequent method of exposure, 100-101, 168
  with complex platinum salts, 122

Ion-exchange technique, for recovery of platinum metals, 13
Iridium
  as hardening agent, 29
  complexes of, 64-65
  compounds of, 53-54
  density of, 39
  effect of heat on, 45
  effect of mineral acids on, 48, 49
  refining of, 11, 13
  retention of, 97-98
  spectrophotometry to determine, 74
  used in radiography, 38
Iridium dioxide, 53
Iridium hexafluoride, 53
Iridium trichloride, 53
Isotopes of platinum metals, 40, 42

Jewelry, platinum metals in, 1, 15, 37

Kidneys
  effect of *cis*-dichlorodiammineplatinum on, 33-34
  effect of platinum metals on, 81, 128

Laboratory equipment, platinum metals used in, 26-27
Laurite, 3
Ligands, nitrogen-donor, 63
Liver
  effect of platinum metals on, 81
  platinum concentration in, 132
Lysogenized bacteria, effect of platinum drugs on, 93

Magnets, from cobalt and platinum, 29
Medical uses of platinum metals, 2, 15, 29-35
Meteorites, platinum metals in, 3
Minerals, platinum metals in, 4-5

Naphtha-reforming, platinum in, 15, 18
Neutron-activation analysis, for determining platinum metals, 75-76, 166
Nickel, recovery of precious metals in refining of, 7, 10
Nitric acid, 138, 164
Nitrogen oxide, 137-38, 182
Nuclear-fuel reprocessing plants, platinum metals in effluent from, 2, 79

Index

Nucleic acid, interaction with platinum complexes, 86–89

Obesity, palladium hydroxide to treat, 29, 103
Occupational exposure to platinum metals, 100, 105, 109, 114, 115
Olefin complexes, 62
Osmium
   alloying behavior of, 44
   catalytic use of, 21
   compounds of, 55–56
   density of, 39
   refining of, 10–11
   retention of, 99
   spectrophotometry to determine, 74
Osmium chlorides, 56
Osmium hexafluoride, 55–56
Osmium tetroxide, 10, 55
   as oxidative catalyst, 22
   formation of, 45, 56
   toxicity of, 102
Oxidation
   ammonia, 19
   platinum-metal resistance to, 14, 166–67
   sulfur, 7
   sulfur dioxide, 19–20
Oxidation states
   of osmium, 55
   of palladium, 50, 57–58
   of platinum, 49, 57–58
   of rhodium, 51–52
   of ruthenium, 55
   stereochemistry and, 57–58

Paint-drying ovens, catalysts used in, 164
Palladium 1, 5
   alloying behavior of, 43–44
   catalytic use of, 20, 21, 166
   compounds of, 50–51
   concentration in air, soil, and water, 126, 127
   effect of mineral acids on, 48
   for relays and switchgear, 23
   in oxidation of ethylene to acetaldehyde, 22
   in synthesis of hydrogen peroxide, 22
   recovery of, 11
   retention of, 95–96
   sales of, 17
   spectrophotometry to determine, 73
   to absorb or desorb hydrogen gas, 35, 37, 46
Palladium chloride
   as germicide, 103
   to treat tuberculosis, 29, 103
   toxicity of, 80, 81
Palladium disulfide, 51
Palladium hydroxide, to treat obesity, 29, 103
Palladium trifluoride, 51
Palladous chloride, 51
Palladous nitrate, 51
Palladous oxide, 50–51
Palladous sulfide, 51
Particles, platinum-metal
   emitted from automobile catalysts, 157–59
   suggested tests of effects of, 178
Passive-transfer tests, 116
Perruthenates, 55
Petroleum. *See* Refining of petroleum
Pharmacologic disposition of platinum metals, 95–100
Platinic chloride, 49
Platinic oxide, 50
Platinic sulfide, 50
Platinosis, 1, 102–3
Platinous chloride, 49
Platinous oxide, 50
Platinum, 1, 5
   alloying behavior of, 43
   catalytic use of, 20, 21
   concentration in air, soil, and water, 126, 127
   effect of mineral acids on, 48
   hardness of, 42–43
   native, 3, 4
   retention of, 96, 97
   sales of, 16
   spectrophotometry to determine, 73
Platinum compounds, 49–50
   allergic reactions to, 105
   prick-test reaction to, 112–14
   toxicity of, 80–82
Platinum fluorides, 50
Platinum metals. *See also* Complexes, platinum; Compounds of platinum metals; Determination of platinum metals; Exposure to platinum met-

als; Properties of platinum metals;
 Refining of platinum metals;
 Separation of platinum metals;
 Therapeutic effects of platinum
 metals; Uses of platinum metals
  availability of, 176, 179
  demand for, 162–63, 172, 176
  dissolved by aqua regia, 67–68
  fusion with base metals, 68
  mutagenic effects of, 102
  reaction with acids, 45–49
  recovery of, 10, 13
  resistance to oxidation, 14, 166–67
  retention of
    by animals, 95–100
    by man, 128
  sales of, 15
  solubility of, 43, 104
  teratogenic effects of, 101–2
Platinum–protein interactions
  allergy tests for, 113–14
  in enzyme inhibition, 84–85
  in x-ray crystallography, 82–84
Platinum salts
  allergic reaction to, 102, 117–19, 164
    bronchial tests for, 119–22
    nasal tests for, 119
  comparison of, 121, 122
  effect on enzyme system, 99
  histamine liberation by, 116–17
  maximal exposure to, 100
  sensitivity to, 106–8
  toxicity of, 80–81
Polarography, for determining platinum
  metals, 76
Potarite, 3
Production of platinum metals, 6, 7–9, 165
Properties of platinum metals, 166–67
  catalytic, 14–25
  chemical, 44–49
  physical, 14, 25–28, 39–44
Proteins, interaction with platinum. *See*
 Platinum–protein interactions
Pyrosulfate fusion, with platinum metals,
  68–69

Quadrivalent metal complexes, 61–62

Radioallergosorbent tests, 115–16
Radiography, iridium used in, 38
Radioruthenium, 79

Recommendations for studies and tests related to platinum metals, 177–79
Recycling of platinum metals, 5, 7, 162–63, 166, 176
Refining of petroleum
  allergic reactions to platinum salts in, 186
  catalytic reforming in, 184, 186–97
  for high-octane unleaded fuel, 182–83
  platinum catalysts used in, 1, 181, 184
Refining of platinum metals
  adverse health effects from, 172
  description of process, 7, 10–13, 166
  platinum concentration in workers, 133
  platinum sensitivity of workers, 109, 114, 115
Relays, palladium used in, 23
Resistors, platinum metals used in, 23–24
Respiratory effects of platinum metals, 100, 103, 105, 106
Rhodium
  alloying behavior of, 44
  as hardening agent, 29
  bactericidal effects of, 90–91
  complexes of, 64–65
  compounds of, 51–53
  effect of mineral acids on, 48
  refining of, 11, 12
  retention of, 96–97
  spectrophotometry to determine, 74
Rhodium chloride, 52
Rhodium salts, 52–53
Rhodium tetrafluoride, 51
Rhodium trifluoride, 52
Rocks, platinum metals in, 5
Rupture diaphragms, platinum used in, 38
Ruthenates, 12, 55
Ruthenium, 1
  as hardening agent, 29
  catalytic use of, 21
  complexes of, 63–64
  compounds of, 54–55
  effect of heat on, 45, 46
  effect of mineral acids on, 48, 49
  isotopes of, 42
  refining of, 11, 12
  retention of, 98–99
  spectrophotometry to determine, 74
Ruthenium dioxide, 54
Ruthenium disulfide, 54
Ruthenium tetrachloride, 55

Index

Ruthenium tetroxide
 as oxidative catalyst, 22
 formation of, 45, 54
Ruthenium trichloride, 54

Sales
 of palladium, 17
 of platinum, 16
 of platinum metals, 15
Sampling of platinum metals, 66–69, 166
Sensitivity to platinum salts, 106, 107–14, 121, 123, 169, 174
Separation of platinum metals
 by chromatography, 70
 by fire assay, 69
 by solvent extraction, 13, 69–70
Silver, refining of, 11, 12
Skin
 allergic reactions to platinum metals, 100, 102–3, 121
 irritation from platinum and palladium compounds, 81, 82, 102
 platinum metals used for treating infections of, 89
 test procedures for, 107–14, 121, 123, 169, 174
Smelters, emission of platinum metals from, 163
Sodium osmate, 12
Soil, concentration of platinum metals in surface, 126, 127, 169
Solvent extraction, for separating platinum metals, 13, 69–70
Sources of platinum metals
 foreign, 5–6, 125, 165
 U.S., 6–7, 126, 165–66
South Africa, platinum metals in, 5, 6
Space heaters, catalysts used in, 164
Spark electrode, platinum alloys used in, 24–25
Spark-source mass spectrometry, to determine platinum metals, 74–75
Spectrophotometry, for determining platinum metals, 73–74
Spectroscopy
 atomic-absorption, 71–72
 emission, 71
 inductively coupled plasma, 73
 x-ray, 3, 74
Sperrylite, 3
Spinnerets, platinum alloys used in, 29

Stereochemistry, of platinum complexes, 82, 83
Stibiopalladinite, 3
Sulfur, oxidation of, 7
Sulfur dioxide oxidation, catalytic use of platinum in, 19–20
Sulfuric acid, for catalytic converters, 151–57, 175–76
Surface analysis, to determine platinum metals, 75
Switchgear, palladium used in, 23

Tetravalent platinum, 80
Tetroxide, volatile, 10–12
Therapeutic effects of platinum metals
 for obesity, 29, 103
 for skin infections, 89
 for tuberculosis, 29, 103
 in arresting tumors, 2, 29–35, 79, 99, 165, 167, 174–75
Thermal reforming, 187–88
Thermocouples, platinum metals used in, 26
Thermodynamics, data on binary alloys, 56
Thermometers, platinum-resistant, 25–26
Tissues, platinum concentration in, 129, 131, 132, 170, 177–78
Toll-refined metals. *See* Recycling of platinum metals
Toxicity of platinum metals and compounds
 dependent on route of administration, 167–68, 173
 dermal irritancies from, 81, 82, 100
 effect on eyes, 81–82, 102
 from inhalation, 100
 histologic changes from, 80
 water solubility and, 173
Tuberculosis, palladium chloride in treatment of, 29
Tubes, platinum used in power and radio, 25

Ultrabasic rocks, platinum metals in, 5
Urine, platinum concentration in, 128, 133, 169
Uses of platinum metals, 1
 as hardening agents, 29
 as histologic stain, 38

based on corrosion resistance, 27–28
catalytic, 2, 14–15, 22–23, 166
  ammonia oxidation, 19–20
  coordination complexes and, 21–22, 49
  dehydrogenation, 21
  exhaust-gas control, 18–19, 134*ff*
  hydrogenation, 20–21
  hydrogen cyanide manufacture, 20
  low-temperature heaters, 23
  sulfur dioxide oxidation, 19–20
dental, 15, 35
electric, 166–67
  contacts for relays and switchgear, 23
  electromechanical electrodes, 24
  fuel cells, 25
  grids for power tubes, 25
  hydrogen products, 25
  resistors and capacitors, 23–24
  spark electrodes, 24–25
for protection of bursting disks, 38
for separation of pure hydrogen, 35, 37

high-temperature
  as flame retardants, 27
  in furnaces, 26
  in gauges, 27
  in laboratory equipment, 26–27
  in thermocouples, 26
  in thermometers, 25–26
in brazing alloys, 37–38
in ferromagnetic alloys, 29
in jewelry, 37
in ornamental surfaces, 37
in pens and phonograph needles, 38
in radiography, 38
in spinnerets and bushings, 28–29
medical, 2, 15, 29–35
U.S.S.R., platinum metals in, 6, 7

X-ray crystallography, 82
X-ray spectroscopy, to determine platinum metals, 3, 74

Zerovalent metal complexes, 58, 59